Preparing Early Childhood Educators to Teach Math

Preparing Early Childhood Educators to Teach Math

Professional Development that Works

edited by

Herbert P. Ginsburg, Ph.D.
Teachers College Columbia University
New York, New York

Marilou Hyson, Ph.D.
University of Pennsylvania
Philadelphia

and

Taniesha A. Woods, Ph.D.
Say Yes Education, Inc.
New York, New York

Baltimore • London • Sydney

Paul H. Brookes Publishing Co.
Post Office Box 10624
Baltimore, MD 21285-0624

www.brookespublishing.com

Copyright © 2014 by Paul H. Brookes Publishing Co., Inc.
All rights reserved.

"Paul H. Brookes Publishing Co." is a registered trademark of
Paul H. Brookes Publishing Co., Inc.

Typeset by Scribe Inc., Philadelphia, Pennsylvania.
Manufactured in the United States of America by
Sheridan Books, Chelsea, Michigan.

Purchasers of *Preparing Early Childhood Educators to Teach Math* are granted permission to download and print the ECME Action Planning Tool for educational purposes. This form may not be reproduced to generate revenue for any program or individual. *Unauthorized use beyond this privilege is prosecutable under federal law.* You will see the copyright protection notice at the bottom of each photocopiable form.

Chapter 3 videos copyright © 2014 Herbert P. Ginsburg, Courtesy of Columbia Center for New Media Teaching & Learning.

Chapter 4 videos copyright © 2014 Juanita Copley.

Library of Congress Cataloging-in-Publication Data

The Library of Congress has cataloged the printed edition as follows:
Preparing early childhood educators to teach math : professional development that works / edited by Herbert P. Ginsburg, Ph.D., Marilou Hyson, Ph.D., and Taniesha A. Woods, Ph.D.
    pages cm
  Includes bibliographical references and index.
  ISBN-13: 978-1-59857-281-0 (pbk. : alk. paper)
  ISBN-10: 1-59857-281-4 (pbk. : alk. paper)
  1. Mathematics teachers—Training of—United States. 2. Early childhood educators—Training of—United States. 3. Mathematics—Study and teaching (Early childhood)—United States 4. Mathematics—Vocational guidance—United States. I. Ginsburg, Herbert, editor of compilation. II. Hyson, Marilou, editor of compilation. III. Woods, Taniesha A., editor of compilation.

QA10.5.P725 2014
372.7'049—dc23

2013050461

2018 2017 2016 2015 2014

10   9   8   7   6   5   4   3   2   1

Contents

About the Editors vii
About the Contributors ix
About the Ancillary Materials xi
Foreword xiii
    Sue Bredekamp, Ph.D.
Acknowledgments xv
Introduction xvii

1   One, Two, Buckle My Shoe: Early Childhood
     Mathematics Education and Teacher Professional Development 1
     Sharon Lynn Kagan and Rebecca Gomez

2   Practices, Knowledge, and
     Beliefs About Professional Development 29
     Marilou Hyson and Taniesha A. Woods

3   Young Children's
     Mathematical Minds: (Almost) All About Ben 53
     Herbert P. Ginsburg

4   Goals for Early Childhood Mathematics Teachers 75
     Juanita Copley

5   General Features of Effective Professional Development:
     Implications for Preparing Early Educators to Teach Mathematics 97
     Martha Zaslow

6   Promising Approaches to Early
     Childhood Mathematics Education Professional
     Development in Preservice Settings: Technology as a Driver 117
     Michael D. Preston

7   Promising Approaches to Early Childhood Mathematics
     Education Professional Development in In-Service Settings 141
     Kimberly Brenneman

8 Evaluating Professional
   Development in Early Childhood Mathematics 173
   Jessica Vick Whittaker and Bridget K. Hamre

9 The Future? 199
   Herbert P. Ginsburg, Taniesha A. Woods, and Marilou Hyson

Appendix: Syllabus for HUDK 4027—
Fall 2013: The Development of Mathematical Thinking 211

Index 217

About the Editors

Herbert P. Ginsburg, Ph.D., Jacob H. Schiff Foundation Professor of Psychology and Education, Department of Human Development, Teachers College, Columbia University

Dr. Ginsburg has conducted basic research on the development of mathematical thinking, with particular attention to young children, disadvantaged populations, and cultural similarities and differences. He has drawn on cognitive developmental research to develop mathematics curricula (Big Math for Little Kids) and storybooks for young children, tests of mathematical thinking, and video workshops to enhance teachers' understanding of students' mathematics learning. He recently developed a model course on early mathematics education for use in colleges and universities. The course makes use of a web-based computer technology (Video Interactions for Teaching and Learning) designed to help prospective teachers improve their craft by making meaningful connections between the cognitive analysis of children's thinking and classroom practice. He has also created computer-based systems (MCLASS: MATH) for helping teachers conduct basic clinical interviews to assess children's mathematical knowledge. In collaboration with colleagues, Dr. Ginsburg is now developing MathemAntics, a computer software to foster young children's (from 3 years to Grade 3) mathematics learning. He holds a B.A. from Harvard University, and his M.S. and Ph.D. are from the University of North Carolina at Chapel Hill

Marilou Hyson, Ph.D., Consultant, Early Childhood Development and Education

Dr. Hyson is also Adjunct Professor in the University of Pennsylvania's Graduate School of Education. Formerly Associate Executive Director and Senior Consultant with the National Association for the Education of Young Children (NAEYC), Marilou co-led the development of the NAEYC/National Council of Teachers of Mathematics position statement on early childhood mathematics, and she prepared a study of teacher preparation for the National Research Council's Committee on Early Childhood Mathematics. Internationally, Dr. Hyson has recently consulted in Vietnam, Indonesia, Bangladesh, and Bhutan through the World Bank and Save the Children, focusing on issues in early childhood professional development. Before joining NAEYC's staff, Marilou was a Society for Research in Child Development (SRCD) Fellow in the U.S. Department of Education as well as Professor and Chair of the University of Delaware's Department of Individual and Family Studies. The former editor in chief of *Early Childhood Research Quarterly,* Dr. Hyson's publications include *Enthusiastic and Engaged Learners: Approaches to Learning in the Early Childhood*

Classroom, The Emotional Development of Young Children, and the forthcoming *The Early Years Matter: Education, Care, and the Well-Being of Children, Birth–8,* coauthored with Heather Biggar Tomlinson.

Taniesha A. Woods, Ph.D., Director of Chapter Programming and Research, Say Yes Education, Inc.

A recurring theme throughout Dr. Woods's career has been the study of issues related to educational equity. Prior to joining Say Yes, Dr. Woods served as a Congressional fellow with the SRCD as part of the American Association for the Advancement of Science on the U.S. Senate Health, Education, Labor, and Pensions Committee, where her portfolio included K–12 and higher education issues. Dr. Woods has also worked as a study director at the National Research Council, where she coedited the 2009 report *Mathematics Learning in Early Childhood: Paths Toward Excellence and Equity.* Most recently and during the writing of this book, Dr. Woods was a senior research associate at Columbia University's National Center for Children in Poverty, where she led action research projects on early childhood education and policy in the areas of mathematics education, teacher professional development, and community-wide education initiatives. Dr. Woods holds Ph.D. and M.A. degrees in developmental psychology from the University of North Carolina at Chapel Hill and a B.A. in psychology and African and African American Studies from the University of Oklahoma.

About the Contributors

Kimberly Brenneman, Ph.D., works at the National Institute for Early Education Research, where she focuses on ways to improve instructional practices that support science and mathematics learning in preschool classrooms. She collaborates on research projects that involve preschool curriculum development, professional development for teachers, and the creation of authentic home–school connections around science, technology, engineering, and math learning. Dr. Brenneman is an author of *Preschool Pathways to Science (PrePS): Facilitating Scientific Ways of Thinking, Talking, Doing, and Understanding* (coauthored with Gelman, Macdonald, & Roman; 2009, Paul H. Brooks Publishing Co.), a guidebook for preschool educators and caregivers, and she is a credited educational consultant for Jim Henson Company's *Sid the Science Kid* television show and web site.

Juanita V. Copley, Ph.D., serves as an early childhood math consultant and currently teaches prekindergarten students and their teachers. She has written and/or edited four books for the National Association for the Education of Young Children (NAEYC) and the National Council of Teachers of Mathematics, served as a consultant for national Head Start, and mentored hundreds of early childhood professionals. As a professor, she coordinated the early childhood program at the University of Houston and was the principal investigator on several grants including a Department of Education grant for 800 early childhood educators in Houston.

Rebecca E. Gomez, Ed.D., is a postdoctoral research fellow at the National Center for Children and Families (NCCF). Her research interests include the study of domestic and international approaches to governance of early childhood education (ECE) systems, the study of professional development systems for ECE, and other topics related to ECE policy and systems building. Prior to joining NCCF in 2009, she worked in various state and national policy settings as an early childhood teacher educator and as an early childhood teacher.

Bridget K. Hamre, Ph.D., is a research associate professor and Associate Director of the University of Virginia's Center for Advanced Study of Teaching and Learning. Dr. Hamre's areas of expertise include student–teacher relationships and classroom processes that promote positive academic and social development for young children; she has authored numerous peer-reviewed manuscripts on these topics. She leads efforts to use the Classroom Assessment Scoring System® (CLASS™) tool as an assessment, accountability, and professional development tool in early childhood and other educational settings. Most recently, Dr. Hamre has engaged in the development and testing of interventions designed to improve the quality of teachers' interactions with students, including MyTeachingPartner™ and an innovative online course for early childhood teachers. Dr. Hamre received her bachelor's degree from the University of California, Berkeley, and her master's and doctorate in clinical and school psychology from the University of Virginia.

Sharon Lynn Kagan, Ed.D., is the Virginia and Leonard Marx Professor of Early Childhood and Family Policy, Codirector of the National Center for Children and Families at Teachers College, Columbia University, and Professor Adjunct at Yale University's Child Study Center. She is also a past president of NAEYC. She is a scholar, pioneer, leader, and advocate. Author of 14 volumes and more than 250 articles, Dr. Kagan has helped shape early childhood practice and policies in the United States and in countries throughout the world.

Michael D. Preston, Ph.D., is Senior Director of Digital Learning for the New York City Department of Education's Office of Postsecondary Readiness. He directs programs that use technology to transform schools, improve teaching and learning, and prepare students for college and careers. He previously worked at the Columbia Center for New Media Teaching and Learning and taught at Teachers College, Columbia University, where he received his Ph.D. in educational psychology.

Jessica Vick Whittaker, Ph.D., is a research assistant professor at the Center for Advanced Study of Teaching and Learning at the University of Virginia. Her research expertise focuses on early childhood classroom processes that promote children's positive academic and social-emotional development. She is currently involved in developing and testing interventions aimed at supporting young children's mathematics, science, and social-emotional skills through the improvement of social and instructional teacher–child interactions. She received her bachelor's in psychology from Duke University and her Ph.D. in human development from the University of Maryland.

Martha Zaslow, Ph.D., is Director of the Office for Policy and Communications of the Society for Research in Child Development (SRCD) and Senior Scholar at Child Trends. As Director of the SRCD Office for Policy and Communications, Dr. Zaslow facilitates the dissemination of research to decision makers and to the broader public and keeps the SRCD membership apprised of social policy and science policy developments related to children and families. She also directs the SRCD Policy Fellowship program, working with the SRCD policy fellows who have placements in the executive branch and Congress. As a senior scholar at Child Trends, Dr. Zaslow conducts research that focuses on professional development of the early childhood work force and approaches to improving the quality of early childhood programs.

About the Ancillary Materials

Purchasers of the book may stream and view the accompanying videos for Chapter 3, *Young Children's Mathematical Minds: (Almost) All About Ben* and Chapter 4, *Goals for Early Childhood Mathematics Teachers*, for educational use. Purchasers may also download, print, and/or photocopy the blank form, ECME Action Planning Tool. These materials are included with the print and e-book and are available at www.brookespublishing.com/downloads with (case sensitive) keycode: 44ns43T1n.

LIST OF ANCILLARY MATERIALS

Video clips for Chapter 3: Young Children's Mathematical Minds: (Almost) All About Ben (Herbert P. Ginsburg)

Video 3.1: Ben with Yellow and Red Chips, Age 3

Video 3.2a: Counting in Farsi, 1–40

Video 3.2b: Counting in Farsi Slowly, 1–19

Video 3.2c: Counting in Farsi Slowly, 1–40

Video 3.2d: Counting in Farsi Faster, 1–40

Video 3.3: Ben Counting, Age 3

Video 3.4: Ben Counting, Age 4

Video 3.5: Ben Playing "Catch My Mistake" Game, Age 3

Video 3.6: Ben Playing "Catch My Mistake" Game, Age 4

Video 3.7: Ben Deciding to Be the Experimental Subject

Video 3.8: Ben Counting from 1 to 29, Age 5

Video 3.9: Ben Counting after 30, Age 5

Video 3.10: Ben Counting to 79, Age 5

Video 3.11: Ben Counting Past 79 with Help, Age 5

Video 3.12: Interviewing to Discuss What Children Know/Teach

Video 3.13: Ben Displaying Variable Command of Enumeration, Age 3

Video 3.14: Ben Counting Bears, Age 3

Video 3.15: Ben Counting Chips, Age 4

Video 3.16: Ben Counting 8 Chips, Age 4

Video 3.17: Elephant Cardinality Problem, Age 3

Video 3.18: Elephant Cardinality Problem, Age 4

Video 3.19: Ben Counting 5 Apples, Age 4
Video 3.20: Ben Counting 3 Apples, Age 4
Video 3.21: Ben Adding Chips
Video 3.22: Ben Adding Pirate Coins
Video 3.23: Ben Counting Chips
Video 3.24: Ben Playing a Game with Janet
Video 3.25: Ben Counting Bears, Age 5

Video clips for Chapter 4: Goals for Early Childhood Mathematics Teachers (Juanita Copley)
Video 4.1: Director Explaining Changing Math Goals
Video 4.2: Teacher Describing Her Understanding of Math
Video 4.3: Teacher Describing His Instructional Games
Video 4.4: Two Girls Playing a Game
Video 4.5: Young Child Making a Block Construction
Video 4.6: Teacher Describing Her Use of Manipulatives
Video 4.7: Teacher Describing Using Math Throughout the Day
Video 4.8: Director Explaining Intentional, Appropriate, and Playful Math

Early Childhood Math Education Action Planning Tool

Foreword

In 2009, the National Research Council (NRC) published a landmark report titled *Mathematics Learning in Early Childhood: Paths Toward Excellence and Equity*. Along with many of the authors in this book, I was privileged to be a member of the authoring committee, which included early childhood mathematics researchers and educators, mathematicians, developmental psychologists, and early childhood specialists. We reviewed the most current research on preschool children's capacities to learn mathematics relative to the quality of early education they typically receive, with particular focus on children whose life experiences limit their opportunities to learn. Sadly, we found that the math achievement gap between children from less economically advantaged backgrounds and their middle-class peers already exists at preschool and is even larger than the literacy gap.

The NRC report identifies the foundational mathematics learning content and skills that predict later success in school as well as evidence-based insights related to curriculum, instruction, and teacher education. The most striking conclusion of the report is that the gulf between what we know and what we are doing for young children is so vast that addressing it calls for a coordinated national initiative. Unfortunately, in the years since the report's publication, little progress has been made to improve either teaching practices or children's learning outcomes, which is why this book is so important and timely.

Arguably, this book is targeted at the most urgent barrier to improving early math education: professional preparation and teacher development. Conversations about program improvement always seem to get caught in a vicious cycle. To improve children's learning, teachers need to do a better job; to improve the quality of teachers' performance, preservice and in-service teacher preparation needs to change. Intervention at this level has the potential to have an impact on entire generations of new teachers and to update the knowledge and practice of the current work force. But on the topic of mathematics, teacher educators need to be educated as well.

In this volume, Ginsburg, Hyson, and Woods tackle this difficult task, drawing on the existing research and identifying questions in need of more study. They incorporate the wisdom of those who have been laboring in the math professional development vineyards for decades and also identify the most recent promising practices. The authors do not shy away from the complexity of the task. Kagan and Gomez lay the groundwork by describing the landscape of the field and identifying the many road blocks that include the diversity of delivery systems and policies as well as the persistent misconceptions about the inappropriateness of teaching mathematics. Hyson and Woods delve deeper into this territory with real-life examples of current practices and the barriers presented by teachers' knowledge, attitudes, and beliefs. Many of the authors emphasize what we now know: Early mathematics learning *is* developmentally appropriate, intellectually engaging, and potentially intensely enjoyable for preschool children.

At the risk of offending the other authors, I must admit that the chapters by Ginsburg and Copley are my favorites. They are rich with examples of real children whose complex mathematics thinking and problem solving come to life for readers via the accompanying online videos. In an enchanting series of vignettes that follow Ben from ages 3 to 6, Ginsburg beautifully models truly effective formative assessment using the clinical interview method. Similarly, Copley demonstrates how she works with a small group of preschoolers and what she has found to be effective through years of implementing professional development with teachers from all background types.

In early childhood education, the lines between preservice and in-service education are blurred. Professional development often takes the place of any preservice preparation. But even when teachers have an early childhood bachelor's degree, they are unlikely to have had relevant coursework or field experience. This book addresses that void by articulating what and how teachers should be learning and even provides a sample syllabus for such a course.

Subsequent chapters address each of the possible intervention points and the supporting research base. Zaslow frames the discussion with what is known about effective professional development in general and its implication for math education. Preston tackles the intransigence of higher education institutions by offering technology-based solutions for preservice education. Brenneman identifies promising practices in professional development, strategies that must be implemented on a large scale to address the needs of the existing work force and current generation of young children. Vick Whittaker and Hamre address the elephant in the room of teacher education: Does it work? How should we evaluate these programs, and what have we learned from existing innovations that are being evaluated?

In their conclusion, Ginsburg and Hyson lay out a blueprint for the future, in effect calling for a different work force than the one we presently have. Achieving this goal will require major investments and policy shifts—a daunting but necessary series of tasks. We now know that young children are more cognitively competent than previously assumed and that they are capable of learning foundational mathematics when provided with effective, developmentally appropriate teaching and learning experiences. The same can be said of their teachers. Just as children need good teachers, their teachers need effective teacher educators. The promise of preschool will never be achieved without well-qualified early childhood professionals at every level. This book challenges us to stop our hand-wringing and get on with it.

Sue Bredekamp, Ph.D.
Early Childhood Education Consultant

Acknowledgments

The findings and recommendations in the 2009 National Research Council (NRC) report *Mathematics Learning in Early Childhood: Paths Toward Excellence and Equity* gave the initial impetus for this book. This landmark report from the NRC panel—a group that included many of this book's contributors—had a clear message: Significant improvements in the professional development of early childhood educators were essential to ensure young children's mathematical confidence and competence. We thank Paul H. Brookes Publishing and especially Astrid Zuckerman for seeing this need and responding to it.

The authors of each chapter—Kimberly Brenneman, Juanita Copley, Rebecca Gomez, Bridget K. Hamre, Sharon Lynn Kagan, Michael D. Preston, Jessica Vick Whittaker, and Martha Zaslow—have our deep appreciation and respect for the expertise, depth of experience, and energy they brought to their task. This commitment went beyond their own chapters; at each stage in developing this book, the editors shared outlines and drafts with the entire group of authors, allowing us to draw on multiple sources of insight as we sought to create chapters that complement and inform one another, with the result that the whole truly is more than the sum of the parts. It was only through authors' openness and generosity that this collaborative editorial process worked so well. It was truly a pleasure to work with this talented, creative group of professionals.

Throughout the book, readers will find rich descriptions of young children's mathematical development and of teachers' and researchers' efforts to understand and promote that development. We acknowledge all these children, teachers, and researchers, and we know that readers of this book will benefit from their examples, both in print and online.

Finally, we acknowledge the diligent assistance of the Brookes editorial and production team, including Sarah Zerofsky, Kari Waters, Susan Hills, and Charlotte Wenger, whose efforts yielded a final product that is both professional and accessible to multiple users.

Introduction

This book serves as an urgent call to action: action needed to develop an infrastructure for the systematic provision and expansion of early childhood mathematics education professional development (ECME PD).

Mathematics is now a top priority for the U.S. education system. Competence in mathematics has an impact on our nation's success at home and in the international arena. However, in the past, math education has been implemented too little and too late, not beginning in earnest until students were in the elementary grades, and often with ineffective curricula and teaching methods. Today, both educational and moral imperatives compel us to start earlier and do better, with developmentally appropriate mathematics education from preschool onward. Mathematics education during early childhood is critical to supporting the "whole child," as high-quality math experiences are important for their own sake and for their contribution to children's competence in other domains such as language, literacy, and social skills. Furthermore, as a nation concerned with educational equity, high-quality math education for young children living in poverty can help to reduce gaps in both opportunity and achievement. With implementation of the K–12 Common Core State Standards in mathematics (National Governors Association Center for Best Practices & Council of Chief State School Officers, 2010), supports must be in place for educators to provide a solid foundation for all children through high-quality early childhood mathematics education (ECME). For this to occur, professional development—the focus of each of the chapters in this book—is essential.

Early childhood educators are responsible for supporting young children's development and learning. Historically, the focus has been on teachers' responsibilities for social, emotional, literacy, and language development—but in recent years, recognition of the importance of ECME has become more prominent. However, as documented in the National Research Council (NRC) report on ECME (2009), prospective and practicing teachers of young children have few and/or inadequate opportunities to receive preparation and training in mathematics education. The gaps in preparation are evident in both "preservice" (i.e., associate, baccalaureate, and masters programs) and "in-service" professional development.

A primary reason that ECME has been neglected is that early childhood educators have not had the essential support and resources necessary to implement high-quality math instruction. Building on the NRC (2009) report, this book focuses on professional development related to math teaching and learning for children from age 3 through Grade 1. Chapters engage readers in research-based discussions of the issues while pointing the way toward practical solutions. Because research shows early childhood professional development can be a primary influence on program quality and improvements in children's learning outcomes in general, we believe professional development has great potential to produce positive change in mathematics specifically.

We hope that this book will be a practical resource for a number of audiences interested in improving professional development in early childhood mathematics:

- College and university faculty in 2- and 4-year teacher educational programs (both generalists in early childhood education and specialists in math education)
- Professional development specialists/trainers working within public school districts, Head Start, child care agencies, and other venues
- Education decision makers such as early childhood education specialists in state departments of education, state prekindergarten coordinators, public school curriculum specialists, and Head Start education directors

The contributors to this book represent a broad range of expertise. Some are experts in early mathematics development and ECME. Others contribute broader expertise in early childhood professional development, and still others enrich the discussion with insight into what is needed to engender systemic changes to ECME through a focus on early childhood systems and early childhood education policy. Throughout, the authors share their own experiences as researchers, teacher educators, designers, and evaluators of innovative approaches to ECME PD.

Together, the nine chapters in *Preparing Early Childhood Educators to Teach Math* provide a set of actionable, research-based strategies for improving ECME PD. The book also provides supplementary resources (available as a download) that professional development providers can use to strengthen the structure and content of their own ECME activities.

In Chapter 1, Sharon Lynn Kagan and Rebecca Gomez set the stage by describing the context in which early childhood education occurs, with a specific focus on math teaching and learning. They begin by describing the characteristics of the early childhood education work force and the varied, often challenging conditions in which practitioners care for and teach young children. The chapter also includes an examination of trends in contemporary early childhood education, including systemic levers for quality improvement, such as state teacher qualification criteria, early learning standards, child assessment requirements, and program quality rating systems. Kagan and Gomez end the chapter with suggestions for changes in mathematics content, interventions, and systems that have the potential to improve ECME—suggestions that are further developed in subsequent chapters.

Chapter 2 takes up this theme by describing the current realities of early childhood instructional practices and professional development opportunities. Marilou Hyson and Taniesha A. Woods describe teaching practices across early childhood settings; the knowledge, beliefs, and attitudes that underlie teachers' instruction; and the professional development available to current and future teachers. When readers finish the chapter they will understand what math-related professional development opportunities are generally available and will be able to compare these opportunities to their own context. Based on the patterns that emerge from examining these realities, Hyson and Woods offer concrete strategies for improving professional development in ECME.

The next chapter turns to the children themselves. Herbert P. Ginsburg's Chapter 3 takes readers on an engaging tour of young children's mathematical development, including their understanding of numbers, written symbols, and concepts of addition and subtraction. The chapter also provides background on children's use of math in their everyday lives and explores the connections among math and other subjects. Readers can watch a child's mathematical development unfold through multiyear videos of one little boy from about the ages of 3 to 5. After reading Ginsburg's chapter, readers will have a deeper knowledge of goals for children's mathematical development and, in turn, how teachers might implement instructional practices that promote children's learning.

Building on Chapter 3's description of goals for children, Juanita Copley uses Chapter 4 to discuss five essential goals for teachers as they strive to implement high-quality early childhood math experiences. These goals range from knowledge of math content to formative assessment to implementation of appropriate teaching practices. Copley includes a brief discussion of the rationale for each goal and the research on which these goals are based. Copley's concrete recommendations will guide readers in planning and implementing their own goals-based ECME PD. Readers will also have an opportunity to watch companion videos that show examples of teachers in action.

In Chapter 5, Martha Zaslow provides a general professional development framework relevant to both the preservice and in-service levels. Drawing on a recent comprehensive literature review for the U.S. Department of Education (2010), Zaslow describes key features of effective early childhood professional development, identifying those that may be especially important in preparing early educators to teach math. In addition, readers will consider the challenges that educators and professional development providers may experience as they aim to improve math teaching and learning outcomes—challenges that readers may be struggling with in their own work.

The next two chapters address more specific math-related professional development issues within preservice and in-service settings. Preservice teacher education has been widely criticized and yet offers great potential to improve mathematics teaching and learning. In Chapter 6, Michael D. Preston provides a thoughtful analysis of current preservice programs and examines new approaches to preservice professional development. Notably, the chapter emphasizes promising practices in the use of technology for educator preparation in early childhood mathematics, illustrated with concrete examples and online resources.

Turning from higher education to in-service professional development, in Chapter 7 Kimberly Brenneman describes effective ECME PD for practicing teachers. The chapter includes a discussion of existing professional development models that show promise for advancing mathematics education in early childhood. The chapter also addresses the practical challenges for implementation of ECME PD within child care, Head Start, prekindergarten, and other settings.

A recurring concern about all professional development has been the lack of rigorous evaluation of its effectiveness. Expanding on the discussions in

Chapters 6 and 7, Jessica Vick Whittaker and Bridget K. Hamre use Chapter 8 to describe the processes and results of evaluations of some existing ECME PD models. Set in the context of a theory of change related to professional development, the chapter includes a constructive critique of current evaluation work and directions for future research. After reading this chapter, readers will not only understand more about the effects of the specific models that are reviewed but will also gain insight into how to effectively use research on ECME PD when considering strategies to bring these models to a larger scale.

Ginsburg, Hyson, and Woods use the final chapter to provide a blueprint for the future of professional development in ECME, including the policies, priorities, and research that will be needed. In Chapter 9, the authors envision new, innovative approaches to professional development in mathematics education for young children—approaches that, although new, are grounded in lessons learned from research and current recommended practices. The chapter addresses the need to recast and broaden the nature of and participants in professional development, with future participants including families, health providers, faith-based communities, and other community members with the potential to create a "math-nurturing culture" for all young children, especially those living in poverty. Ginsburg, Woods, and Hyson also argue that achieving this vision will require creative uses of technology as well as a radically different way of thinking about the targets and methods of professional development.

Many of the book's chapters refer to and give examples of innovative delivery methods for ECME PD, especially new uses of technology. Consistent with this emphasis, the book's streaming and downloadable content as well as the e-book version of the book supplement the traditional print format so that practitioners and policy makers have access to comprehensive, user-friendly information that can immediately be applied in real world settings. These supplementary resources include illustrative videos and a higher education course syllabus—all of which allow those using this book to see mathematics teaching and learning in action so that they are better equipped to apply material in the book to practice.

Our goal in writing this book and providing the related technologies is to provide the field of early childhood education with an actionable vision for high-quality mathematics teaching and learning with professional development as the lever for change. We also know that President Barack Obama's call for improved access to and quality of preschool programs presents an opportunity and challenge that the field must be ready to take on—in math no less than in other areas, as research demonstrates that mathematics is a cornerstone for young children's foundational skills and knowledge.

The timing for this book could not be more critical—these issues need to be addressed and the time is now. The following chapters provide the deep insight and concrete strategies necessary to help educators to teach high-quality mathematics and to help all children get off to a strong start in school and in life.

REFERENCES

National Governors Association Center for Best Practices & Council of Chief State School Officers. (2010). *Common Core State Standards.* Washington, DC: Authors. Retrieved from http://www.corestandards.org

National Research Council. (2009). *Mathematics learning in early childhood: Paths toward excellence and equity.* Committee on Early Childhood Mathematics, C.T. Cross, T.A. Woods, & H. Schweingruber (Eds.). Center for Education, Division of Behavioral and Social Sciences and Education. Washington, DC: National Academies Press.

U.S. Department of Education, Policy and Program Studies Service. (2010). *Toward the identification of features of effective professional development for early childhood educators: Literature review.* Washington, DC: Author.

*To the memory of Jane Knitzer, a staunch advocate
for children, especially the poor and poorly served. She was especially
concerned with early education and would have applauded the efforts
described in this book to promote the effective professional development
of those who seek to educate young children, our most precious resource.*

1

One, Two, Buckle My Shoe

*Early Childhood Mathematics
Education and Teacher Professional Development*

Sharon Lynn Kagan and Rebecca Gomez

From the beginning of a child's life, he or she is exposed to mathematics. From a child's Apgar score to the litany of nursery rhymes that creep into the cadences of a young child's experiences, numbers abound. In some cases, they appear with great intentionality early on (e.g., the Apgar score); in other cases, they are so natural that they are barely noticed (e.g., the routine chanting of nursery rhymes). Who has not asked a young child her age, only to be demurely, and sometimes silently, greeted by two or three little fingers popping up? Who, irrespective of continent or culture, has not almost instinctively helped a young child count or even match numbers with pebbles or pennies? Whether explicitly or implicitly transmitted, most adults value mathematics and mathematical concepts for very young children.

As children mature into adulthood, the mathematical stakes become higher. Writing in a recent edition of *Education Next,* Vigdor notes, "In the 21st century workplace, mathematical capability is a key determinant of productivity" (Vigdor, 2013). Substantiating the point, he indicates that college graduates who majored in subjects related to math, engineering, and the physical sciences earn an average of 19% more than those who specialize in other fields. Indeed, precollegiate mathematic Scholastic Assessment Test (SAT) scores predict higher earnings among adults, whereas verbal SAT scores do not. These startling facts, coupled with the finding that American students routinely fall behind their peers on international comparisons of mathematical achievement, have spiked interest in educational and economic circles about the importance of mathematics instruction and achievement to the well-being of the country and its economy. As a result, a bevy of funding for science, technology, engineering, and math projects is burgeoning in the public and private sectors, with innovative efforts being launched from preprimary to postsecondary education.

Lest we think such a focus on mathematics is new, a quick review of educational efforts suggests the precise opposite. A fundamental cornerstone of schooling since the Common School, emphasis on mathematics seems to increase in times of social crises: During World War II, concern about mathematics abounded as soldiers were unable to calculate the simple and necessary trajectory of artillery shells; the launching of Sputnik evoked scores of programs and policies focusing on ginning up American students' capacities in mathematics, the sciences, and foreign languages; and even the report that awoke this nation to the need for educational restructuring, "A Nation at Risk"(National Commission on Excellence in Education, 1983), pointed out the disparaging fact that only a small fraction of high school students managed to complete calculus. With these realizations came a litany of efforts to "re" mathematics: to reconceptualize, refocus, reinvent, or reform it.

It is important to note that, while many of the efforts to improve children's performance in mathematics focused at the elementary, secondary, and postsecondary levels, there was also a tremendous, but often unnoticed, focus on mathematics that took hold for young children before even entering the academic arena. For example, "One, Two, Buckle My Shoe" was first published in *Songs for the Nursery* in London in 1805 as a counting song. More formal mathematics efforts also appeared throughout the history of preprimary education. They took the form of Friedrich Froebel's "gifts," Maria Montessori's number rods, and the Sheffield math program, clearly indicating that mathematics was not simply the purview of formal schooling and federal policy. Stated simply, although school reform efforts captured the policy allure, schools could not claim eminent domain over mathematics instruction. Transcending the ambience of early care and education, mathematics and mathematical processes have been considered by early educators for decades, if not centuries.

But early mathematics has faced a number of critical challenges. Foremost among them are deep-seated attitudinal barriers, with some emanating outside of early education and others sourced within the field itself. Those outside the early childhood field historically have dismissed the overall potency of early childhood education, equating it with mere babysitting or aimless play. Recently, although these general sentiments have abated so that the early years and play are now well-regarded, misconceptions toward mathematics persist. Characterized by a sentiment of hopelessness, it sadly suggests that "not all children are equally prepared to embark on a rigorous math curriculum on the first day of kindergarten, and there are no realistic policy alternatives to change this simple fact" (Vigdor, 2013). Whether lamenting deep-seated social and economic inequities, the lack of adequate teachers, or outdated curricula to move children along a potent mathematical trajectory, there are concerns about the viability of contemporary early mathematics to prepare all children for the future.

Such concerns also exist within the field of early childhood education; they vary in orientation but not in magnitude. Today, many early educators are concerned with the overacademicization of the field. Rife with angst that young children will be denied a childhood, that too much focus on "content" areas will depress children's innate curiosity and creativity, and that externally imposed and standardized curricula and assessments will deplete a commitment to individualization and diversity, some reject the emerging emphasis on content areas and processes, including mathematics. Echoing this perspective, with its focus on the *re*production rather than the *pro*duction of knowledge (Moss, 2013), an overemphasis on didactics defies both the canons and the traditions of early education. Others suggest that the divide between so-called academic and developmental orientations is a false dichotomy. This stance suggests that rich material provides the content while productive play is the process of early education; these merge naturally both to fuse the dichotomy and to provide exciting and meaningful learning experiences for young children. Whatever one's disposition on these stances, there can be little doubt that such debate characterizes the discussion about early education in general, which in turn frames attitudes toward and conversations about early mathematics.

Less often spoken about, and that which this volume seeks to address directly, is the continuing concern about the adequacy of mathematics preparation and the experiences of those who work with young children. Certification requirements are limited, as are opportunities for professional development in early education in general, much less in early mathematics. With a growing literature that confirms the importance of the teachers' role in early education, far more prominence has been given to certification and professional development in general, but too little of both pertain to teachers of early mathematics. Moreover, many early educators who have very limited certification and professional development are somewhat limited in their own mathematical abilities; often, such personal fears translate into less time being devoted to mathematics and less frequent opportunities for mathematics to enter the curriculum. These inhibitors to making mathematics a rich, meaningful component of early education are often detrimental to children's early learning and development, catapulting the discourse on early mathematics education and professional preparation to new heights.

To begin addressing these issues, this chapter staunchly advocates the "One, Two, Buckle My Shoe" theme. It, similar to the nursery rhyme written 2 centuries ago, acknowledges that numbers and mathematics matter; they matter a great deal. Furthermore, this chapter urges us to metaphorically "buckle [our] shoes." In each round, the famed nursery rhyme begs for action: "Five, six, pick up sticks / Seven, eight, lay them straight." With that same proactivity and to that end, we intend to "pick up [the] sticks" by addressing the context for early mathematics. We do so in two sections. The section titled "The General Context for Change" presents the general status of early education, first by delineating the somewhat complicated nature of the early childhood work force and then by discussing current trends in the context of contemporary early education. Following this broad contextual review, in "Mathematics for Young Children in Context," we focus particularly on early mathematics. Then, building on these contextual reviews, in "Conclusion: The Content and Process of Change," we look ahead and delineate critical issues as well as suggestions for moving forward to inspire new approaches to early childhood mathematics professional preparation and development. In doing so, we seek to tackle head-on the trenchant issues (i.e., the "sticks") and "lay them straight." Our goal in this work, then, is to buckle up and buckle down to the hard work of understanding the general early childhood context, the context for early mathematics, and the steps that might be considered as a productive agenda to advance early childhood mathematics professional development.

THE GENERAL CONTEXT FOR CHANGE

Efforts that seek to hasten social reform cannot be successful without having a clear understanding of the context in which that reform exits. Propelling early childhood mathematics forward, then, demands an understanding of the general early childhood context in which reform of early mathematics is squarely placed and by which it is heavily contoured. Blending historical context, demographic information, and emerging professional development efforts, this general contextual discussion is focused on two points. First, we present an overview of the status and preparation of the early childhood work force. Second, in characterizing the current early childhood context, we cite several trends, selecting those that are most pertinent and likely to have an impact on the reform of early childhood mathematics professional preparation.

The Early Childhood Work Force and Its Preparation: Roots and Realities

Written about in scores of publications, the early childhood work force can only be characterized as diverse and complex. In part, its complexity exists because of diversity on nearly every measureable variable (e.g., degree requirements, delivery systems, credentialing options, certification requirements). In this section, we explain that diversity by presenting the field's roots. We then examine the realities of the current state of the early childhood work force. We present the demographics of the work force, examine several dimensions that characterize its reality, and conclude by discussing the nature of early childhood certification and credentialing.

Roots: The Historical Evolution of Early Childhood in the United States

The contemporary status of the early childhood work force has deep historical roots in the importance of family in American society. Characterized by privacy and primacy, the historical American ethos frames sentiments that regard public support to young children and families as an intrusion into the sanctuary of the family. Families are expected to be responsible for their children, and only when families falter should the government step in to provide support. Reflecting public reluctance to support early childhood education (ECE) outside the home and lacking a firm public commitment, services for all young children in the United States grew episodically and inconsistently, varying from locale to locale. Some services emerged in the public sector, but more often than not, service delivery expanded in the private sector, often in privately owned settings (e.g., privately owned and licensed child care centers or family child care homes) or in informal care arrangements (e.g., family, friends, neighbors). Given this ethos, it is not surprising that ECE in the United States evolved as a field without a coherent set of requirements for teacher preparation and certification. The ripple effects of the lack of attention to professional development for the ECE work force affect every aspect of teaching and learning, not the least of which is early childhood mathematics.

Reality 1: Demographics

Echoing the legacy of ECE programmatic diversity, the early childhood teaching work force is also diverse in its composition and even in how it defines itself. Although we use the term "teacher" to refer to adults who are working with young children across a variety of settings, the U.S. Bureau of Labor Statistics (BLS), for example, uses different terms to categorize ECE teachers depending on the setting in which they work. Adults who work with children from birth to kindergarten entry are called either *child care workers* or *preschool teachers* (U.S. Bureau of Labor Statistics, 2011a). Child care workers, according to BLS, are individuals who are paid to care for children in private households, family child care homes, center-based child care programs, and after-school programs. Preschool teachers are defined as individuals working in similar settings, but with a more explicit focus on their role as educators, and typically working with children ages 3–5 years. The BLS definition of preschool teachers also includes individuals who are working in publicly funded prekindergarten programs (U.S. Bureau of Labor Statistics, 2011a). Individuals who are employed as child care workers and preschool teachers total about 1.75 million individuals; approximately 1.3 million individuals are classified as child care workers, and another 475,000 are classified as preschool teachers (U.S. Bureau of Labor Statistics, 2011b). The challenge with these figures is that they do not take into account the number of individuals working within informal care arrangements. Although the number of individuals employed within this segment of the ECE work force is impossible to document exactly, various estimates suggest that an additional 550,000 individuals provide care and educative experiences for young children in informal settings (Fowler, Bloom, Talan, Beneke, & Kelton, 2008; Karolak, 2008). This number brings the figure for the birth-to-5 segment of the ECE teaching work force to approximately 2.3 million (Ryan & Whitebook, 2012)—a very large group indeed. In addition to those teaching children in preprimary settings, the ECE work force is also composed of teachers of children in kindergarten and Grades 1–3. Though more difficult to disaggregate this group of individuals from the statistics on elementary teachers as a whole, BLS notes that in 2012 there were 164,910 individuals classified as *kindergarten teachers* in the United States, and an additional 1,415,590 *elementary school teachers* (U.S. Bureau of Labor Statistics, 2012a), about half of which likely comprise teachers of Grades 1–3.

Irrespective of the differences in how these individuals are defined and classified, ECE teachers share a number of important commonalities. First, the early childhood teaching work force has historically been, and continues to be, primarily female. At last estimate, more than 95% of the ECE work force was female (Ryan & Whitebook, 2012), reflecting historical gender patterns (Kagan, Kauerz, & Tarrant, 2008; Karolak, 2008). Second, ECE teachers are predominantly white (Ryan & Whitebook, 2012), though this varies depending on the overall racial and ethnic composition of the locale in which

ECE programs exist. In California, for example, the ethnic composition of the early childhood teaching work force includes more minorities than is typical nationally, closely mirroring the overall demographics of the state (Kagan, Gomez, & Friedlander, 2012). A number of studies have found that family child care providers and assistant teachers are more diverse than are teachers in center-based child care programs (Chase, Moore, Pierce, & Arnold, 2007; Marshall, Dennehy, Johnson-Staub, & Robeson, 2005; Ryan & Whitebook, 2012). A third commonality is the fact that ECE teachers tend to be in their late 30s to early 40s. Recent data (Ryan & Whitebook, 2012) show that the average age of an ECE teacher is 39 years, with teachers working in public prekindergarten settings tending to be slightly younger.

Reality 2: Compensation

Across the board, ECE teachers are poorly compensated for their work. Compensation includes both wages and benefits (e.g., health benefits, pensions). The BLS calculates the average hourly wage for a child care worker at $9.26 per hour, or $19,300 per year, and the average wage for preschool teachers at $12.35 an hour, or $25,700 per year (U.S. Bureau of Labor Statistics, 2012a, 2012b). These figures have been consistent over the past few years (U.S. Bureau of Labor Statistics 2010a, 2010b, 2011a, 2011b, 2012a, 2012b), illustrating that persistently low wages are an endemic problem for the ECE teaching work force. Compounding the problem of low wages is the fact that only about 33% of ECE teachers receive full health benefits (Whitebook et al., 2006). The lack of health care coverage, coupled with low wages, often forces teachers to leave the field after relatively short careers.

Reality 3: Instability

Low compensation is a problem unto itself. It is, however, closely linked to high rates of work force turnover and instability. The reasons for such turnover are many. For example, among those who have the necessary qualifications, many are drawn to the higher-compensating early childhood jobs in the public school system. For other individuals, the allure of a higher-paying job in another field may entice them to leave the ECE field altogether.

Moreover, some members of the early childhood work force leave because they find the demands associated with increased professional development onerous. In some cases, for example, teachers are asked or required to participate in professional development without financial and programmatic support. In effect, this means that already underpaid teachers are expected to do more work without short-term support or any long-term assurance of increased compensation. In an attempt to make quality improvements, such demands can evoke a heavy burden on an already fragile work force. Recognizing the situation, outstanding efforts to support and compensate ECE teachers for their professional development are emerging. One example is the Teacher Education and Compensation Helps (T.E.A.C.H.) program, wherein teachers who enroll in a degree program are provided time during the workday to complete assignments and school-related tasks. When the teacher earns her degree, she receives an increase in pay; in return, she commits to teaching in the program for a certain contract period (e.g., 12 or 24 months). Initiatives such as T.E.A.C.H. have helped bolster the stability and qualifications of the ECE work force, but they are not as widespread as they need to be in order to have a large-scale impact on the work force. Some states are also thinking about stability in terms of the availability of career advancement for teachers; teachers may be more motivated to stay in a position or in the field if they understand that there are opportunities to advance and know the steps to pursue them over time. To this end, many states have developed career lattices that delineate the qualifications needed as individuals progress through the system.

Reality 4: Teacher Preparation Requirements

Whereas the work force is quite homogenous with regard to demographics, levels of compensation, and relative instability, it differs dramatically with regard to the requirements for teacher qualifications and preparation. Here, diversity is the norm, resulting in underdevelopment of the field. Degree requirements for an ECE teacher vary dramatically with regard to 1) the state he or she lives in, 2) the type of program in which he or she works, and 3) the teaching position he or she holds (e.g., ages of children, lead or assistant teacher). To explain this diversity, we analyze qualifications by program type: child care centers, family child care homes, Head Start, and public prekindergarten; we also discuss disparities in qualifications between teachers of infants and toddlers and teachers of preschool and prekindergarten.

Child Care Centers As we have noted, the vast majority of center-based child care programs function within the private sector. Although these programs are often required to be licensed by the state in which they operate and are thus monitored to ensure they meet basic health and safety requirements, only a few states extend such oversight to the qualifications teachers must obtain before they can enter the classroom. For example, only a few states mandate that teachers have some sort of formal preservice training. This preservice experience can range from taking college courses (as is required in California) to holding a nationally recognized credential such as the Child Development Associate (CDA) credential (as is required in Hawaii, New Jersey, Illinois, Minnesota, and Vermont), to completing a vocational program (as is required in Colorado, Massachusetts, New Hampshire, and Delaware; National Education Association, 2010).

Family Child Care Homes Overall, the requirements for ECE teachers working in family child care settings are lower than for those working in center-based programs. As of 2010, only 24 states had articulated preservice requirements for individuals working in a family child care homes, some of which included the requirement for teachers to hold a national credential, such as the CDA (National Child Care Information Center, 2010), though as of 2008, only about 3% of family child care providers had obtained CDA certification (Kagan et al., 2008). In addition, 38 states do require some sort of in-service professional development—usually taking the form of workshops rather than credit-bearing professional development (National Child Care Information

Center, 2010). Again, these requirements do not stipulate that teachers participate in professional development related to early childhood mathematics.

Head Start   In contrast to the privatized nature of child care centers and family child care homes, Head Start is a federally funded and monitored program for children ages 3–5 from low-income families. Head Start programs are subject to requirements set forth by the federal government, including teacher preparation requirements. Federal regulations established by the Improving Head Start for School Readiness Act of 2007 (Head Start Act of 2007; PL 110-134) mandate that by 2013, every Head Start classroom will have at least one teacher 1) who has an associate's, bachelor's, or graduate degree in ECE; 2) who has an associate's or bachelor's degree in any field and coursework equivalent to an ECE major, with experience teaching preschool- age children; or bachelor's degree in any field and coursework equivalent to an ECE major, with experience teaching preschool-age children; or 3) who is a member of Teach for America (an alternative certification program that trains and places new teachers), has passed the Praxis exam, and is participating in ongoing professional development offered by Teach for America (Administration for Children and Families, 2013). In addition, by September 2013, at least half of Head Start teachers nationally were required to have 1) a bachelor's or advanced degree in ECE or 2) a bachelor's or advanced degree in any subject along with coursework equivalent to an ECE major and experience teaching preschool-age children (Administration for Children and Families, 2013).

Public Prekindergarten   Publicly funded prekindergarten programs exist in 38 states (Barnett & Whitebook, 2011). Among these states, there is wide variation in teacher requirements for those working in public pre-K classrooms. Twenty-six of these thirty-eight states have defined the minimum level of qualification for prekindergarten teachers as being a bachelor's degree with ECE content (Barnett & Whitebook, 2011). The remaining states require that teachers have some type of preservice professional development in ECE, such as an associate's degree or a CDA credential. Although these requirements mean that prekindergarten teachers do, overall, have higher levels of educational attainment than their counterparts in child care centers and family child care homes, there is still a great deal of variation among states in the requirements for prekindergarten teachers.

Age Group: Infant and Toddler Teachers   We have defined the ECE work force as consisting of teachers of children from birth to Grade 3. Although the focus of this book is primarily teachers of children age 3 through Grade 1, we highlight teachers of infants and toddlers here because of the importance of children's learning in the first 3 years of life. Given this reality, professional development for teachers of infants and toddlers is particularly critical, yet requirements for teachers of this age group are lower than those for their counterparts teaching children ages 3–5. Few states require any sort of credential or degree as a prerequisite to working with infants and toddlers (National Education Association, 2010); as such, there are fewer available professional development supports for this subset of the ECE work force. Although there is an infant/toddler version of the CDA credential, and although some states such as Pennsylvania and New Hampshire offer infant and toddler credentials, these specialized professional development requirements are episodic and typically only required for those who are participating in a publicly funded quality improvement initiative, such as a Quality Rating and Improvement System (QRIS) or rated license. Content, instead, favors social-emotional development, language acquisition, and physical development, often at the exclusion of mathematics. The challenge for this segment of the work force, then, has historically been and continues to be twofold: 1) extremely low baselines for qualifications and 2) limited supports for professional development for all segments of the work force, including, but not limited to, the teaching of early childhood math.

Reality 5: Professional Preparation Programs

Just as diversity characterizes the requirements for ECE certification, so too do the institutions that provide the professional preparation for teachers vary. An estimated 1,349 institutions of higher education (IHEs) in the United States offer an early childhood degree program of some kind; this represents approximately 30% of all IHEs in the country (Maxwell, Lim, & Early, 2006). As states invest in professional development systems and actively work to build the capacity of the IHEs in those states to offer degree programs in ECE, this percentage is expected to rise. Within this number, as of early 2013, more than 30 states have at least one college with an NAEYC-accredited associate's degree program in ECE, with some states having more than 10 (e.g., North Carolina with 20 and South Carolina with 14; National Association for the Education of Young Children [NAEYC], n.d.). There are also more than 450 nationally-recognized B.A. and graduate degree programs in ECE in 38 states (NAEYC, n.d.). The conundrum at present, however, is that, despite the availability of these professional preparation programs, the lack of consistent state requirements for early childhood teacher preparation renders professional preparation programs unnecessary for many who work with young children outside of public school settings.

Reality 6: Voluntary Credentials

In an attempt to fill the vacuum created by such inconsistencies in teacher requirements, several national organizations have propagated voluntary ECE credentials, two of which are highlighted here: the CDA and the National Board for Professional Teaching Standards (NBPTS).

Recently revised and perhaps the most widely recognized credential in the early childhood field, the CDA is a competency-based credential that is available to any adult in the early childhood work force. It is offered by the Council for Professional Recognition, which was established for the explicit purpose of advancing the CDA as a potential solution to the underdeveloped nature of the ECE work force. As an entry-level credential, the CDA is viewed as a "key stepping stone on the path of career advancement in ECE" (Council for Professional Recognition, 2011, p. 1), and for those teachers who pursue the CDA, it is often a precursor to entry into a degree program. The competency areas in which candidates must demonstrate proficiency include safe and healthy learning environments, children's physical and intellectual competency, children's social-emotional development, relationships with families, program management, and professionalism. In demonstrating proficiency, candidates must develop a portfolio that provides documentation on how they have met the competency areas. The portfolios also include documentation on how they will address children's mathematical learning (Council for Professional Recognition, 2013).

The second credential available to teachers is the NBPTS ECE credential. One of many credentials offered by the NBPTS, the ECE credential is available to teachers who have earned a bachelor's degree and want the portability of a nationally recognized credential combined with widespread recognition (2012). The NBPTS ECE generalist standards ask teachers to focus on fostering cognitive development, including a number of mathematical processing skills. There is not, however, a set of standards dedicated to mathematics (NBPTS, 2012). The major challenge of the viability of this credential for the ECE work force is the fact that, as we have already noted, few teachers have a bachelor's degree and thus are unable to obtain this credential.

As states move forward in building the infrastructure for professional development for the ECE work force, these credentials may serve as important pillars on which to build a solid foundation. At present, however, they are utilized unevenly within and across states, adding to the complex set of realities of the ECE work force.

Trends in the Broader Contemporary Early Childhood Context

Given these realities of the ECE work force, it is also important to understand the major trends that frame the contemporary early childhood work force. Three trends (popularization, accountability, and quality enhancement) are discussed next; these are couched within the sentiment that the current context is one of opportunity, despite the present fiscal limitations and uncertainties.

The Trend Toward Popularization

Without doubt, the past decade has witnessed a surge in the understanding of, and support for, early education worldwide. Understandings related to the importance of early education have occurred because of the popularization of unequivocal research attesting to its potency as a social and educational intervention that produces short- and long-term benefits. Though the actual research had been mounting for decades, it was not until it was translated and popularized to business and policy leaders that its importance became manifest in policy. Dating to the end of the 1990s and the first decade of the 2000s, ECE has moved front and center as a policy issue, with programs and services burgeoning throughout the world. The United Nations Educational, Scientific and Cultural Organization (UNESCO; 2008, p. 50) notes that from 1999 to 2006, access to preschool facilities for children from 3 years of age increased from 112 million to 139 million, with the global preprimary gross enrollment rate (GER) averaging 79% in developed countries and 36% in developing countries (though enrollment in the United States is markedly lower than both of these percentages, as we note in the paragraph that follows). Australia is a good case in point, with its services to young children increasing from 1,078,710 in 2006–2007 to 1,158,690 in 2009–2010 (Organisation for Economic Co-operation and Development, 2012a, p. 332). The Economist Intelligence Unit, in its 2012 report "Starting Well," noted that, of the 45 countries surveyed, 33 countries provided access to services to more than 50% of the children in the country (Economist Intelligence Unit, 2012). Notably, these changes have reached previously unserved populations in Africa, Southeast Asia, and remote parts of the Indian subcontinent.

Domestically, such changes have also taken vigorous hold. Preschool enrollments are on the rise, with a majority of young children enrolled in some type of formal educative experience—approximately 69% of 4-year-olds, as noted in a recent report (Organisation for Economic Co-operation and Development, 2012b). but public investments in early education have also increased overall, with investments of nearly $5 billion being made across all U.S. states (Barnett, Carolan, Fitzgerald, & Squires, 2011; Silberman, 2012). In addition, considerable investments are being made by the federal government in Head Start (more than $7 billion in 2011), the Child Care and Development Fund (more than $5 billion in 2012), and supports for children with disabling conditions through the Individuals with Disabilities Education Act (IDEA [PL 108-446]; nearly $4 million in 2011) (Administration for Children and Families, 2012; U.S. Department of Education, 2012).

Moreover, throughout the United States, scores of organizations have adopted ECE as their action priority, with each mounting numerous programs and innovations. Business organizations, including the Business Roundtable, are investing considerable effort in early education. Politicians are routinely mentioning early education as a part of their election platforms, and major organizations with broad agendas, including the National Governors Association, the Council of Chief State School Officers, and the National Commission for Education and the Economy, are all advancing early education. In addition, in early 2013 the Obama administration announced a major initiative in which the federal government would partner with states to fund high-quality preschool for all young children. In short, the zeitgeist for young children has edged forward significantly, precipitating a change in the framing questions. In the past, the dominant question focused on the importance and value of early education; with answers to these questions well established, now the prevailing question has turned to issues of delivery and prioritizations, asking, how and in what ways should services be expanded?

The Trend Toward Enhanced Policy/System Building

With the trend toward increased services has come the clear knowledge that past policy apparatuses, with their scattershot and uncoordinated approaches, will no longer suffice. Today's efforts have turned from a focus on supporting new programs to developing an early childhood system, replete with mechanisms that will provide durable funding and durable governance of the early childhood field. In part, the push for such an orientation stems from decades of experience that has witnessed "here today, gone tomorrow" programs—programs that were the cherished jewel of one politician and have been summarily eradicated as another political leader came to office. In part, the push for a systems approach is the realization that not only are the administrative costs associated with mounting so many programs in diverse agencies excessive and inefficient, but diverse funding streams proffering indefinite amounts of support each year leave the field vulnerable and disorganized. Finally, there has been some concern that, whereas 3- and 4-year-olds have been the subject of much policy attention, children younger than age 3 have been somewhat marginalized in the policy arena. A systemic focus, which implies a focus on children from birth to age 8, helps refocus policy attention on all young children traditionally included in the early childhood age spectrum.

Efforts to build early childhood systems have taken many forms, with new and important policies and programs foremost among them. The advent of the Early Learning Challenge Fund, a competitive federal grant program, has reinforced the focus on developing and supporting elements of the early childhood system, offering states considerable incentives to advance their systems of assessment and accountability, and quality enhancement schemes. Moreover, systemic efforts are taking hold in many states, with new approaches to consolidated governance appearing in Massachusetts, Maryland, Pennsylvania, and Washington, among others. Inherent in the addition of a systems-based orientation is the understanding that policy support for programs alone, although necessary, is insufficient to render the quality services that result in solid results in children's development and performance. Quality programs must be augmented by durable funding and policies that yield a quality infrastructure that includes a firm commitment to professional development.

The Trend Toward Increased Program Quality

Having already mentioned the need for and trend toward quality, it is important to underscore that quality enhancement efforts are taking many forms. Three are discussed here. The first addresses efforts to improve overall accountability, including efforts to specify and align standards, develop effective assessment tools, and collect and report data with efficiency. The second addresses the trend toward producing increased continuity among the diverse settings that have an impact on children's development, notably the linkages between home/community and schooling, including the push for a greater emphasis on dual-language learners as well as the focus on strengthening linkages between preprimary and primary settings. Finally, there is a clear growth in the understanding of the importance of program improvement as a quality elixir.

Accountability, Standards, Assessment, and Data Traditionally in the early childhood field, accountability has focused on the nature and quality of the programs children received. This orientation was premised on the fact that children enrolled in higher-quality settings tended to demonstrate better outcomes, with three factors predicting such quality: group size, child-to-adult ratios, and teacher experience and training. As the field has matured, two things have occurred to expand this thinking. First, with more data, it has become clear that other variables are also important predicators of children's outcomes, including the prevalence of an organized approach to the learning environment, the engagement of parents, and the specific nature of teacher–child interactions. This changing definition of quality has been accompanied by more sophisticated data that clearly suggest that all children in high-quality programs do not make comparable progress. Children's presenting variables (e.g., their physical and mental health, the nature of their home settings, their mothers' education level) predict some variation in outcomes. As a result, and presaged by growing budgetary and business concerns about educational efficacy in general, new approaches to accountability that focus more directly on child outcomes have taken hold in ECE.

Such a focus on children's development and performance has evoked a call for the international and domestic development of early learning and development standards (ELDS), often called guidelines or benchmarks. Though the terms differ, all states have such expectations for their 3- and 4-year-old children, and the majority of states also have such documents for infants and toddlers. ELDS are seen not simply as accountability tools but also as mechanisms for improving the overall quality and consistency of children's early experiences and more equitable child expectations and outcomes. In some cases, standards are being used as the basis for curriculum development and revision, professional development, and the development of comprehensive assessments that will address all domains of children's development. Closely aligned with the development of standards is a growing attention to redirecting assessment so that it takes into consideration the unique learning styles, needs, and capacities of young children. For example, through the Early Learning Challenge Fund, a number of states, such as Ohio and Maryland, are in the process of creating new kindergarten entry assessments; other states are looking at inventive ways to calibrate existing assessments with one another and with their state's standards. Whatever approach is being considered or used, there is much effort being expended to devise and collect information on young children's developmental and educational progress, which can be used to guide both policy and practice.

Aligned with these initiatives, states are embarking on significant efforts to create or update their data management systems. In some states, these efforts are being merged with those underway in primary and secondary education. Ambitiously, some states are conceptualizing their data systems to include the multiple state agencies whose work touches young children: for example, physical health, mental health, social services, and education. Often, these efforts begin with the provision of child identification numbers that enable the tracking of services received and outcomes produced throughout children's educational experiences. Advancing this work, the Early Childhood Data Collaborative has developed key elements of a data system, such as a unique statewide child identifier; child-level demographics and program participation information; child-level data on development; the ability to link child-level data with K–12 and other key data systems; a unique program-site identifier with the ability to link children and the ECE work force; program-site structural and quality information; a unique ECE work force identifier with the ability to link with program sites and children; individual-level data on ECE work force demographics, education, and professional development information; a state governance body to manage data collection and use; and transparent privacy protection and security policies and practices (Early Childhood Data Collaborative, 2012).

Finally, many of the efforts in this quality-enhancement area are shrouded with controversy. Not all early educators accept the need for standards, assessments, or accountability, fearing that delineating such expectations will place too much pressure on young children. Kindergarten entry assessments, for example, are now becoming the norm in many states, yet they remain contested by many in the field. Indeed, formidable attitudinal challenges join the technical challenges associated with standards development and alignment and test construction as well as the structural challenges that accompany cross-agency and cross-state collaborations.

Continuity Across Settings Less controversial but no less complex, new efforts to develop continuity across settings are gaining currency. Long-standing commitments to families and communities have characterized early education since its inception, with the early existence of parent cooperatives and the parent engagement policy statements that are the fiber of the nation's Head Start program. Today, given the increasing diversity of families coming into early education settings coupled with the increasing needs of all parents, emphasis has been placed on enhancing parenting education, support, and empowerment efforts. In some cases, these have taken the form of devoting increased funds to parent support, mounting new programs that are designed to meet the needs and schedules of families, and efforts that meet the needs of dual-language families.

Efforts to promote continuity across the settings that serve young children are also burgeoning, with increased efforts to link preprimary and primary education. Such efforts appear under an umbrella of terms (e.g., birth to age 3, ages 3–5, pre-K–Grade 3, pre-K to age 20, transition, alignment), but all are devoted to enhancing the alignment of programs, services, professional development, standards, assessments, and/or curricula. That this work has become so prolific has also raised concern among some scholars and practitioners who are worried about the "schoolification" of ECE (Moss, 2013); others note schoolification as a possibility but seek to discern and advance the positive benefits of reforming preprimary and primary education to render them more child centered as a result of this linkage (Bennett, 2013; Kagan, 2013).

Program Improvement Although there are many efforts to enhance program quality, only a few are discussed here. The most widespread quality improvement program is the QRIS. QRIS efforts vary in scope and magnitude, but they share a commitment to engaging personnel in the assessment of program quality against a prescribed standard of quality. The results of these ratings are then verified by external raters, and stars or indicators of quality are accorded and made public. In some states, the QRISs are linked to increased levels of fiscal reimbursement, with the amount increasing with the program's quality ranking or number of stars.

Other efforts to enhance program quality include program accreditation and the related Accreditation Facilitation Projects (National Association for the Education of Young Children, 2014). These efforts seek to engage programs in a review of their services and provide supports for quality enhancement. It should be noted that most of the program quality efforts are being linked to some form of professional development already discussed.

MATHEMATICS FOR YOUNG CHILDREN IN CONTEXT

The previous section, "The General Context for Change," focused on the general context of ECE in the United States as a prelude to the discussion of the status of early childhood mathematics specifically. In this section, we turn our focus to early childhood mathematics by beginning with a discussion of the status of work force qualifications and preparation related to early childhood mathematics, followed by a discussion of the current trends relevant to early childhood mathematics in the section "Trends in Contemporary Early Childhood Mathematics."

The Early Childhood Work Force and Mathematics Preparation

Mathematics Requirements for Early Childhood Education Teachers: A Needle in the Haystack?

Much like looking for a needle in a haystack, the degree to which the early childhood work force has exposure to early childhood mathematics by way of meeting teacher preparation requirements is difficult to discern. The variability of teacher preparation requirements makes it difficult to tease out what, if any, mathematics content is mandated as part the preparation requirements for teachers working in a specific program within a particular state. Indeed, in conducting our research for this chapter, we found only one program type that expressly articulated that teachers complete professional development in early childhood mathematics content: Head Start.

As a part of the Head Start Act of 2007, the federal government put forth teacher preparation requirements that require professional development specific to each of the domains of learning in early childhood, including mathematics. The guidance offered by the Administration for Children and Families notes that "coursework equivalent to a major relating to early childhood education includes but is not limited to courses that focus on mathematics, as well as other domains of learning" (Administration for Children and Families, 2013, p. 1).

Although it is possible that states may follow Head Start's leadership in this arena, presently, there appears to be one constant among the wide variation in teacher preparation requirements: There is little in the way of requirements that mandate that teachers must have professional preparation and development that includes content specific to early childhood math.

Early Mathematics Beliefs and Preferences: Implications for Teaching and Teacher Professional Development

Challenges to mathematics professional development not only result from the lack of requirements for formal preparation, they are compounded by complex beliefs and preferences that pertain to early mathematics. As explained in Chapter 2, which discusses teacher beliefs and preferences about professional development in early childhood mathematics, research suggests that teachers of young children are not as invested in the importance of mathematics as they are in other domains of learning (National Research Council, 2009; Varol, Farran, Bilbrey, Vorhous, & Hofer, 2012). Prevailing belief systems among some

ECE teachers are based on the assumption that young children are neither interested in nor capable of higher levels of cognitive functioning at an early age. Although this view is changing, research about teacher beliefs suggests that the importance of mathematics is not yet fully recognized in ECE settings.

Early educators also strongly believe (as they are trained to) in a holistic approach to development. Focusing on any one domain in favor of others tilts learning and privileges a given domain—something that is deemed undesirable. Indeed, most early childhood teachers are not prepared to teach domain-specific activities (Brenneman, Stevenson-Boyd, & Frede, 2009). In short, teacher beliefs, preferences, and prevailing instructional practices are challenges that must be addressed as early childhood mathematics professional development is advanced.

Trends in Contemporary Early Childhood Mathematics

Despite the challenge presented, several important efforts to advance early mathematics are gaining currency. As a prelude to discussing recommendations for action, we review three in this section: the trend toward acknowledging the importance of early mathematics; the trend toward increased research; and the trend toward improved practice.

The Trend Toward Acknowledging the Importance of Early Mathematics

Much like the growing attention that is being accorded to early childhood in general, early childhood mathematics is gaining increased acknowledgment. In part, this attention comes as a residual benefit of increased public and private commitments to young children and their families in general. Enhanced attention to early mathematics also emanates from the trend toward accountability discussed in the section "The Trend Toward Popularization." With its focus on child/student outcomes and with the specification of mathematics standards for students in Grades K–12 in the Common Core State Standards, the push for results in mathematics is soaring and promoting attention to early mathematics. Correspondingly, the push to create a learning continuum from birth through the early years of schooling renders both language arts and mathematics increasingly important to early childhood and preprimary education.

The movement at the primary level, however, is but one precipitating factor accelerating the importance of early mathematics. Within the early childhood field itself, interest in early childhood mathematics is soaring. Position statements from the field are noting the importance of integrating early mathematics with all domains (NAEYC & National Council of Teachers of Mathematics [NCTM], 2002). There has been a call to "provide for children's deep and sustained interaction with key mathematical ideas" (NAEYC & NCTM, 2002), clearly demonstrating a commitment to fostering early mathematics. This document has been supported by a 2006 effort by the National Council of Teachers of Mathematics (NCTM) and by the refinement of mathematics standards in Head Start. In short, early mathematics is receiving increased attention from diverse sectors in ways that are more explicit and more direct than ever before.

The Trend Toward More Research and Research Syntheses

Some of the increased acknowledgment of the importance of early math stems from our increasing knowledge of its importance to young children's development and learning. Much of this information emanates from a growing research base. Well synthesized in the National Research Council's report *Mathematics Learning in Early Childhood* (2009), the research found that although virtually all young children have the capacity, and most have the interest, to be meaningfully involved with mathematics, the potential to learn mathematics is not universally realized. The report notes that opportunities to learn mathematics are not routinely prevalent, particularly for economically disadvantaged children. Several reasons are given: Mathematics is deemed a less important discipline than literacy; there is limited high-quality mathematics instruction for young children; and mathematics, when presented in early childhood classrooms, tends to be embedded in the broader curriculum rather than treated as a separate subject and is secondary to other learning goals. Recognizing this paucity, the seminal report makes nine recommendations that call for a nationally coordinated early childhood mathematics initiative that would position early mathematics far more prominently on the agenda of policy makers, teacher educators, teachers, and families.

Important for its thoroughness and objectivity, *Mathematics Learning in Early Childhood* also contributed greatly to moving the early mathematics agenda forward in two additional ways. First, it synthesized the existing research, and in doing so, pointed out important research gaps that need attention. Second, because of the importance of the National Academy of Sciences and the quality of the report's scholarship, early childhood mathematics rose to increased prominence. Throughout the nation, early mathematics was increasingly seen as essential to a quality early education; moreover, as a historically neglected domain in early childhood, often being ignored in favor of social and emotional development and literacy, mathematics gained new legitimacy as an essential domain of inquiry.

Another synthesis document, *Math Matters* (Schoenfeld & Stipek, 2011), presented convincing data amassed by scholars that attests to the importance of early math. Citing Duncan and colleagues (2007), *Math Matters* notes that when comparing math, literacy, and social-emotional development at kindergarten entry, "early math concepts such as knowledge of numbers and ordinality were the most powerful predictors of later learning." *Math Matters* notes that these findings are consistent with those of other scholars (e.g., Grissmer, 2011; Romano, Kohen, Babchishin, & Pagani, 2010). In addition, a growing body of research indicates that early mathematical proficiency is associated with later proficiency in not only mathematics but reading as well (Duncan et al., 2007; Schoenfeld & Stipek, 2011) and may even be linked to increased rates of high school graduation. *Math Matters* goes on to suggest, "Although the mechanisms underlying such associations are not yet understood, the importance of early math—and thus of access to it for all students—is clear" (Schoenfeld & Stipek, 2011, p. 3). The report presents information on curricular and pedagogical approaches to math as well as cogent reviews of professional preparation for early mathematics teachers. Concluding with a call to action that embraces policy and pedagogy, the document both chronicles and advocates for a revitalized commitment to mathematics for young children.

Along with these influential research syntheses, individual scholars long interested in mathematics for young children continue to produce work that makes clear the cognitive foundations for early mathematics learning. This work addresses features of developing mathematical constructs and knowledge, critical changes in the development of mathematical understandings, and the relationship between mathematical development and the development of cognitive skills in other domains of learning. Scholars are also exploring how the acquisition of mathematical knowledge and skills may vary from child to child, with a special focus on addressing the unique ways in which dual-language learners and children with multiple risks or developmental delays acquire and process mathematical concepts. An especially fertile line of inquiry relates to examining children's learning trajectories as a way to better understand the precise steps and pathways children must traverse as they become proficient in one level before moving on to the next (Sarama & Clements, 2009).

Moreover, and pertinent to this volume, the body of research on professional development for teachers of early mathematics is growing as well. Neuman and Kamil (2010) have devoted a volume to the issue of professional development for ECE teachers and report on seven studies that address early childhood mathematics interventions and their positive impacts on children. Collectively, the studies reveal positive child outcomes, particularly in reducing disparities among the performances of children from different socioeconomic backgrounds. Central to the findings was the professional development received by teachers. Most specifically, gains in students' math learning have been related to teachers' spontaneous discussions about math concepts (Klibanoff, Levine, Huttenlocher, Vasilyeva, & Hedges, 2006). Varol and colleagues (2012) examined preschool teachers' math instruction and their participation in a preschool math intervention program. The results suggest that there is a strong relationship between ongoing professional development paired with in-classroom supports (i.e., coaching and mentoring) and the quality of teachers' math instruction, with teachers with the weakest skills often deriving the greatest benefit from the professional development. Clearly, though limited in number, these examples suggest that a robust scholarly base related to early mathematics teaching and learning is emerging and that productive intellectual debates are occurring; in turn, this supports the importance of research as a seminal pillar in advancing national interest in, and concern for, early mathematics.

The Trend Toward Improved Practice

The improvement of practice is taking many forms, two of which are addressed here: the trend toward improved standards and the trend toward improved tools. At the outset, it should be noted that each of these areas of improved practice in ECE are also prevalent in primary and secondary education in the United States. Moreover, each of the areas is not without its fair share of controversy, at both the preprimary, and primary and secondary levels. Well beyond the scope of this analysis to present and discuss both the K–12 efforts and the controversies associated with each, suffice it to say that controversy, although generally acknowledged as negative, can multiply the attention given to an issue and can infuse invention with creativity and activity. Such is the case with early mathematics and the standards and tools being developed to enhance its practice.

Improved Standards As noted, not long ago, early educators associated standards with program specifications or requirements, often associated with health, safety, and the educational environment. In recent years, with the advent of accountability, standards in ECE have also come to be thought of as expectations for what young children should know and be able to do. Now prevalent in all 50 states, standards emanate from intentions to improve both the quality of instruction and the equity of children's opportunities to gain access to quality instruction. Generally, early learning and development standards embrace five developmental domains, including physical health and motor development, social and emotional development, language and literacy development, cognitive development, and approaches to learning, although standards are not always grouped according to these categories or domains. Regarding this discussion, one might ask where mathematics fits in and how well represented are mathematics standards in the states' ELDS. To address these questions, an analysis of the 50 state standards, plus the District of Columbia and select territories, was undertaken, with the results indicating that "mathematics standards were not only terribly inconsistent across the states, but were also somewhat insufficient in number and content" (Scott-Little, Kagan, Reid, & Castillo, 2011, p. 23). The study revealed, for example, that one state had three mathematics standards for 4-year-olds, whereas another state had 193 indicators. Moreover, the data revealed a heavy emphasis on the number and operations areas, with far less focus on mathematical processes, including problem solving and estimation. Although disturbing at first glance, the analysis suggested specific areas for improvement and also realerted those concerned about curriculum and teacher preparation that grave inconsistencies existed in what was being expected from young children in terms of their mathematical development.

At the national level, noteworthy efforts to develop standards or guidelines in mathematics have been undertaken by NCTM, dating back to 1989 when their standards for elementary and secondary school mathematics included kindergarten standards. These were revised in 2000 when prekindergarten was included in the prekindergarten to Grade 2 age band. This pioneering effort was followed up in 2002 with a position statement jointly developed by the National Association for the Education of Young Children (NAEYC) that was designed to advance policy and practice related to early mathematics. Later, in 2006, NCTM released *Curriculum Focal Points for Prekindergarten Through Grade 8 Mathematics,* which described the most significant mathematical concepts and skills at each grade level. In addition, the Head Start Child Development and Early Learning Framework provides important national standards related to children's early mathematics development, as do the Common Core State Standards for students in kindergarten through Grade 12. In summary, then, the trend toward the development and refinement of early mathematics standards emanates from diverse levels and agencies of government and from the private sector. Together, they confirm an increased commitment to specifying what young children should know and be able to do in the area of mathematics.

New Tools: Curricula and Technology If inconsistency characterizes standards as one quality formula, diversity characterizes the second group of quality enhancement efforts, notably curriculum and technology. At the outset, such diversity is expressed in terms of what actually constitutes curriculum. On the one hand, curriculum in ECE is often included across content areas or domains. To make learning meaningful for young children, this integration calls for the linkages of two or more disciplines in a single activity or learning experience (Schickedanz, 2008). In this approach, a broad topic of interest to children (e.g., animals, families, weather) is typically selected, and then learning activities that incorporate literacy, numeracy, the arts, and science are blended together so that the children are not conscious of learning a specific discipline. In this case, mathematics is a secondary objective. Alternatively, other approaches to curriculum development suggest that, although necessary, an inclusive approach to mathematics is not sufficient. Rather, an inclusive curriculum must be supported by a focused curriculum that advances mathematics as its primary goal, a recommendation that is also discussed at length in Chapter 7. This stance is supported by data that reflect that children spend a good deal of their time in prekindergarten engaged in routine activities and meals and snacks, with only very limited time spent on mathematics; in fact, a study by the National Center for Early Development and Learning found that, on average, children were engaged in mathematics activities for only 6% of the entire day (Early et al., 2005).

Examples of curricula that significantly address mathematics abound (e.g., Big Math for Little Kids, Building Blocks, Core Knowledge, Creative Curriculum, High/Scope, Pre-K mathematics, and Number Worlds). A great deal of research suggests that such focused curricula have the *potential to* increase the mathematics achievement of children living in poverty (Campbell & Silver, 1999; Clements & Lewis, 2009; Griffin, 2004), but limited data address the actual efficacy of specific mathematics curricula. What evidence exists, notably from Building Blocks and Big Math for Little Kids, suggests that the mathematical knowledge of children from low-income communities does increase significantly when exposed to these curricula (Clements & Lewis, 2009; Clements & Sarama, 2007, 2008). It is important to note that structured curricula are being regarded with new optimism as a means of enhancing early mathematics outcomes for young children. Here, the idea of quality is closely aligned with outcomes.

Other criteria for quality exist, including the ability to engage children meaningfully in productive learning experiences. To that end, and supporting the reality that young children are excited by hands-on learning experiences, various kinds of mathematics manipulatives have long been a part of early education. Cuisenaire rods and blocks are notable examples. In the contemporary world, such hands-on manipulatives are being joined by technological manipulatives that arguably offer more flexibility than their noncomputer counterparts (Clements & McMillen, 1996). Beyond hands-on manipulatives, the use of computers and other technologies offers children the opportunity to review their work, to reflect on why decisions were made, and in some cases, to enhance their capacity to carry out dexterously complex activities, such as drawing complex or symmetrical shapes. Moreover, the use of computers may more specifically, the preparation of early mathematics teachers. Some topics for such consideration might include: Under what conditions, and for whom, does what kind and dosage of early mathematics matter? To what end, and under what conditions and with what intensity, should early mathematics be trained in teachers? What is the most productive role of new technologies in advancing a mathematics agenda? What are the professional and public dispositions toward early mathematics and how do they vary across populations? Hardly complete, this list will no doubt be richly augmented by the contributions of this volume.

Having only a single study, although helpful, does not meet the conditions for a "solid knowledge base." Multiple studies are often needed to represent diverse population segments, to examine nuanced findings, and ultimately, to confirm and/or negate a body of findings. As a result, the need to develop research is reliant on having qualified researchers to carry out the work. Attention and incentives must be provided to encourage talented and sophisticated researchers to build the solid knowledge base.

A Social Strategy

Armed with a solid knowledge base, a carefully crafted social strategy can be devised. Such a strategy must consider the context carefully, as this chapter suggests. It must also use policy tools to advance its work, including legislation, executive orders, and regulations in the public sector. It must also include a clear plan of action for engaging private philanthropy, foundations, and the media. Equally important, such a strategy must take into consideration the changing nature of higher education in this country. With its movement toward privatization and online degrees, careful thought must be given to the blend of efforts and educational formats/mechanisms that will create a well-prepared work force. Such a strategy must also be certain to account for those already in the early childhood work force as well as for those preparing to enter. It must provide for changes in the American demographics and for variations in pedagogical approaches to early learning and early mathematics. Less difficult than in the past, developing social strategies has become an industry replete with individuals specializing in moving reform agendas via social planning and marketing.

Public Will

Fostering change in the way early mathematics preparation is conceptualized and popularized is a crucial element of generating public will. As with the development of a social strategy, there are experts to carry out this work, including those with expertise in social networking and web-based media strategies. They, however, will need guidance from those with deep knowledge in early mathematics. Building on a convincing knowledge base and incorporating a social strategy, the development of public will often benefits from presentation of the negative consequences that build up as a result of avoidance of the issue. Beyond sharing the consequences of issue avoidance, the media are increasingly adept at telling and communicating key stories. Capitalizing on a public event (e.g., leveraging the 2012 school shooting incident in Newtown as a platform for debates about gun control and the second amendment), the media has been influential in building interest in a cause. How to make early childhood mathematics preparation mediaworthy in a digital era is a critical challenge for individuals with media expertise. Commandeering those individuals with care and speed should be considered a priority. This volume seeks to launch the nation on such a strategy path. By consolidating the literature and by hearing from diverse voices, the volume may provoke interest in the issue. The challenge for those concerned with the preparation of early childhood mathematics teachers is not to stop there. Understanding the general and the mathematics early childhood context, offering some critical issues, and providing an approach to think about social change has been the content of the chapter; propelling "One, Two, Buckle My Shoe" action toward social change has been its goal.

FOR REFLECTION AND ACTION

1. Kagan and Gomez discuss what often seems to be an aversion to math on the part of early childhood educators because teaching math is not seen as developmentally appropriate or because math is too "academic" for young children. Have you seen evidence of these attitudes? If so, what are your thoughts about how to change them?

2. You may or may not already be familiar with the characteristics of the early childhood work force. From the authors' description, which characteristics do you think may be potential obstacles to the improvement of educator effectiveness in teaching mathematics, and which may be potential strengths or supports?

3. An important point in this chapter is that state-level requirements and policies may influence the content of, and resources available for, early childhood professional development in the field of mathematics. To learn more, look at some of the resources in this book's appendix, the Chapter 7 appendix, or other resources available to you. What is the situation in your state, and how does this affect your math-related professional development opportunities? For example, what are the teacher certification requirements in your state related to mathematics, and how are those reflected in your institution's requirements?

4. The authors discuss at some length the influential 2009 report on early childhood mathematics from the National Research Council. Various findings from this report are also referred to in most of the other chapters in this book. If you are not already familiar with the report, you might read it either in print or online (http://www.nap.edu/catalog.php?record_id=12519) and focus on those issues that are most relevant to your current or future work.

5. If you work in the policy sector or are planning a policy-oriented career, you might want to explore further some of the policy levers identified by Kagan and Gomez that potentially support improved professional development in early childhood mathematics. You might especially consider which of these is most likely to have an impact on your state, district, or community setting.

REFERENCES

Administration for Children and Families. (2012). *Statutory degree and credentialing requirements for Head Start teaching staff* (ACF-IM-HS-08-12). Retrieved from http://eclkc.ohs.acf.hhs.gov/hslc/standards/ims/2008/resour_ime_012_0081908.html

Administration for Children and Families. (2013). *Head Start Act: Fact sheet*. Washington, DC: Author, Office of Head Start. Retrieved from http://eclkc.ohs.acf.hhs.gov/hslc/standards/Head%20Start%20Act

Barnett, W.S., Carolan, M.E., Fitzgerald, J., & Squires, J.H. (2011). *The state of preschool 2011: State preschool yearbook*. New Brunswick, NJ: National Institute for Early Education Research.

Barnett, W.S., & Whitebook, M. (2011). *Degrees in context: Asking the right questions about preparing skilled and effective teachers of young children*. New Brunswick, NJ: National Institute for Early Education Research.

Bennett, J. (2013). A response from the coauthor of a strong and equal partnership. In P. Moss (Ed.), *Early childhood and compulsory education: Reconceptualizing the relationship* (pp. 52–71). London, England: Routledge.

Brenneman, K., Stevenson-Boyd, J., & Frede, E.C. (2009, March). *Math and science in preschool: Policies and practice* (Preschool Policy Brief, Issue No. 19). New Brunswick, NJ: National Institute for Early Education Research. Retrieved from http://nieer.org/resources/policybriefs/20.pdf

Campbell, P.F., & Silver, E.A. (1999). *Teaching and learning mathematics in poor communities*. Reston, VA: National Council of Teachers of Mathematics.

Chase, R., Moore, C., Pierce, S., & Arnold, J. (2007). *Child care workforce in Minnesota: 2006 statewide study of demographics, training and professional development*. Report prepared for the Minnesota Department of Human Services. Saint Paul, MN: Wilder Research.

Clements, D.H., & McMillen, S. (1996). Rethinking "concrete" manipulatives. *Teaching Children Mathematics, 2*(5), 270–279.

Clements, D.H., & Sarama, J. (2007). Effects of a preschool mathematics curriculum: Summative research on the Building Blocks project. *Journal for Research in Mathematics Education, 38*(2), 136–163.

Clements, D.H., & Sarama, J. (2008). Experimental evaluation of the effects of a research-based preschool mathematics curriculum. *American Educational Research Journal, 45*(2), 443–494.

Clements, P., & Lewis, A.E. (2009, April). *The effectiveness of the Big Math for Little Kids curriculum: Does it make a difference?* Paper presented at the American Educational Research Association Annual Meeting, San Diego, CA.

Council for Professional Recognition. (2011). *About the Child Development Associate (CDA) Credential*. Washington, DC: Author. Retrieved from http://www.cdacouncil.org/the-cda-credential/about-the-cda

Council for Professional Recognition. (2013). *Child Development Associate 2.0: Transition guide*. Washington, DC: Author.

Duncan, G.J., Dowsett, C.J., Claessens, A., Magnuson, K., Huston, A.C., Klebanov, P., . . . Japel, C. (2007). School readiness and later achievement. *Developmental Psychology, 43*(6), 1428–1446.

Early Childhood Data Collaborative. (2012, September). *Developing coordinated longitudinal early childhood data systems: Trends and opportunities in Race to the Top Early Learning Challenge applications*. Retrieved from http://www.ecedata.org/files/ECDC-RTTT-Sept27%20%281%29.pdf

Early, D., Barbarin, O., Bryant, D., Burchinal, M., Chang, F., Clifford, R., . . . Barnett, W.S. (2005). *Pre-kindergarten in eleven states: NCEDL's multi-state study of pre-kindergarten and study of State-Wide Early Education Programs (SWEEP)*. Preliminary descriptive report. Chapel Hill, NC: Frank Porter Graham Child Development Institute.

Economist Intelligence Unit. (2012). *Starting well: Benchmarking early education across the world*. Retrieved from http://www.managementthinking.eiu.com/sites/default/files/downloads/Starting%20Well.pdf

Fowler, S., Bloom, P.J., Talan, T.N., Beneke, S., & Kelton, R. (2008). *Who's caring for the kids? The status of the early childhood workforce Illinois—2008.* Wheeling, IL: McCormick Center for Early Childhood Leadership, National Louis University.

Griffin, S. (2004). Number Worlds: A research-based mathematics program for young children. In D.H. Clements, J. Sarama, & A.-M. DiBiase (Eds.), *Engaging young children in mathematics: Standards for early childhood mathematics education* (pp. 325–342). Mahwah, NJ: Lawrence Erlbaum Associates.

Grissmer, D. (2011, October). *Rethinking the importance of early developmental and academic skills in predicting long term achievement and achievement gaps.* Paper presented at a meeting of the National Science Foundation REESE Principal Investigators, Washington, DC.

Kagan, S.L. (2013). David, Goliath and the ephemeral parachute: The relationship from a United States perspective. In P. Moss (Ed.), *Early childhood and compulsory education: Reconceptualizing the relationship* (pp. 130–148). London, England: Routledge.

Kagan, S.L., Gomez, R.E., & Friedlander, J.W. (2012). The status of early care and education teacher preparation in the United States. In W.E. Fthenakis (Ed.), *In Natur-Wissen schaffen.* Bremen, Germany: University of Bremen.

Kagan, S.L., Kauerz, K., & Tarrant, K. (2008). *Early care and education teaching workforce at the fulcrum: An agenda for reform.* New York, NY: Teachers College Press.

Karolak, E.J. (2008). *Investing in early education: Paths to improving children's success: Hearing before the Committee on Education and Labor, House of Representatives,* 110th Cong., 2nd Sess.

Klibanoff, R.S., Levine, S.C., Huttenlocher, J., Vasilyeva, M., & Hedges, L.V. (2006). Preschool children's mathematical knowledge: The effect of teacher "math talk." *Developmental Psychology, 42*(1), 59–69.

Marshall, N.L., Dennehy, J., Johnson-Staub, C., & Robeson, W.W. (2005). *Characteristics of the current early education and care workforce serving 3–5 year-olds* (Massachusetts Capacity Study Research Brief). Wellesley, MA: Center for Research on Women, Wellesley College. Retrieved from http://www.wcwonline.org/earlycare/workforcefindings2005.pdf

Maxwell, K.L., Lim, C.-I., & Early, D.M. (2006). *Early childhood teacher preparation programs in the United States: National report.* Chapel Hill: University of North Carolina, Frank Porter Graham (FPG) Child Development Institute.

Moss, P. (2013). The relationship between early childhood and compulsory education: A properly political question. In P. Moss (Ed.), *Early childhood and compulsory education: Reconceptualizing the relationship* (pp. 2–49). London, England: Routledge.

National Association for the Education of Young Children. (2014). *Accreditation facilitation projects.* Washington, DC: Author. Retrieved from http://www.naeyc.org/academy/primary/localsupport/afp

National Association for the Education of Young Children. (n.d.). *Accredited early childhood programs.* Washington, DC: Author. Retrieved from http://www.naeyc.org/ecada/ecada_programs

National Association for the Education of Young Children & Fred Rogers Center. (2012). *Technology and interactive media as tools in early childhood programs serving children from birth through age 8.* Position Statement. Washington, DC: Author.

National Association for the Education of Young Children & National Council of Teachers of Mathematics. (2002). *Early childhood mathematics: Promoting good beginnings.* Washington, DC: Author. Retrieved from http://www.naeyc.org/files/naeyc/file/positions/psmath.pdf

National Board for Professional Teaching Standards. (2012). *Early childhood generalist: Assessment at a glance.* Washington, DC: Author.

National Child Care Information Center. (2010). *Minimum preservice qualifications, orientation/initial licensure, and annual ongoing training hours for family child care providers in 2008.* Retrieved from http://nccic.acf.hhs.gov/pubs/cclicensingreq/cclr-famcare.html

National Commission on Excellence in Education. (1983). *A nation at risk.* Washington, DC: U.S. Department of Education.

National Council of Teachers of Mathematics. (2006). *Curriculum focal points for prekindergarten through grade 8 mathematics: A quest for coherence.* Reston, VA: Author.

National Education Association. (2010). *Raising the standards for early childhood professionals will lead to better outcomes.* Washington, DC: NEA Policy and Practice Unit, Center for Great Public Schools.

National Research Council. (2009). *Mathematics learning in early childhood: Paths toward excellence and equity.* Committee on Early Childhood Mathematics, C.T. Cross, T.A. Woods, & H. Schweingruber (Eds.). Center for Education, Division of Behavioral and Social Sciences and Education. Washington, DC: National Academies Press.

Neuman, S.B., & Kamil, M.L. (2010). *Preparing teachers for the early childhood classroom: Proven models and key principles.* Baltimore, MD: Paul H. Brookes Publishing Co.

Organisation for Economic Co-operation and Development. (2012a). *Starting strong III: A quality toolbox for early childhood education and care.* Paris, France: OECD Publishing.

Organisation for Economic Co-operation and Development. (2012b). *United States—Country Note—Education at a glance 2012: OECD indicators.* Paris, France: OECD Publishing..

Richmond, J.B., & Kotelchuck, M. (1984). Commentary on changed lives. In J.R. Berrueta-Clement, L.J. Schweinhart, W.S. Barnett, A.S. Epstein, & D.P. Weikart (Eds.), *Changed lives: The effects of the Perry Preschool Program on youths through age 12* (pp. 204–210). Ypsilanti, MI: High/Scope Educational Research Foundation.

Romano, E., Kohen, D., Babchishin, L., & Pagani, L.S. (2010). School readiness and later achievement: Replication and extension study using a nation-wide Canadian survey. *Developmental Psychology, 46*(5), 995–1007.

Ryan, N., & Whitebook, M. (2012). More than teachers: The early care and education workforce. In R. Pianta, W.S. Barnett, L.M. Justice, & S.M. Sheridan (Eds.), *Handbook of early childhood* (pp. 92–110). New York, NY: Guilford Press.

Sarama, J., & Clements, D. (2009). *Early childhood mathematics education research: Learning trajectories for young children.* New York, NY: Routledge.

Scott-Little, C., Kagan, S.L., Reid, J.L., & Castillo, E. (2011, December). *Early mathematics standards in the United States: Understanding their content.* Prepared for the Heising-Simons Foundation. New York, NY: National Center for Children and Families.

Schickedanz, J.A. (2008). *Increasing the power of instruction: Integration of language, literacy, and math across the preschool day.* Washington, DC: National Association for the Education of Young Children.

Schoenfeld, A.H., & Stipek, D. (2011). *Math matters: Children's mathematical journeys start early.* Retrieved from http://earlymath.org/earlymath/wp-content/uploads/2012/03/MathMattersReport.pdf

Silberman, S. (2012). *Early childhood education: Investment not expense.* Bethesda, MD: Education Week.

United Nations Educational, Scientific and Cultural Organization. (2008). *Overcoming inequality: Why governance matters.* Education for All Global Monitoring Report 2009. Oxford, England: Oxford University Press.

U.S. Bureau of Labor Statistics. (2010a). *Occupational employment and wages, May 2010: Childcare workers.* Retrieved from http://www.bls.gov/oes/current/oes399011.htm

U.S. Bureau of Labor Statistics. (2010b). *Occupational employment and wages, May 2010: Preschool teachers, except special education.* Retrieved from http://www.bls.gov/oes/current/oes252011.htm

U.S. Bureau of Labor Statistics. (2011a). *Occupational outlook handbook, 2010–2011 edition: Child care workers.* Retrieved from http://www.bls.gov/oco/ocos170.htm

U.S. Bureau of Labor Statistics. (2011b). *Occupational outlook handbook, 2010–2011 edition: Teachers—preschool, except special education.* Retrieved from http://www.bls.gov/oco/ocos317.htm

U.S. Bureau of Labor Statistics. (2012a). *Occupational outlook handbook, 2012–2013 edition: Childcare workers.* Retrieved from http://www.bls.gov/ooh/personal-care-and-service/childcare-workers.htm

U.S. Bureau of Labor Statistics. (2012b). *Occupational outlook handbook, 2012–2013 edition: Preschool teachers.* Retrieved from http://www.bls.gov/ooh/Education-Training-and-Library/Preschool-teachers.htm

U.S. Department of Education. (2012). *Preschool grants for children with disabilities.* Retrieved from http://www2.ed.gov/programs/oseppsg/funding.html

Varol, F., Farran, D.C., Bilbrey, C., Vorhaus, E.A., & Hofer, K.G. (2012). Improving mathematics instruction for early childhood teachers: Professional development components that work. *NHSA Dialog: A Research-to-Practice Journal for the Early Childhood Field, 15*(1), 24–40.

Vigdor, J. (2013). Solving America's math problem. *Education Next, 13*(1). Retrieved from http://educationnext.org/solving-america%E2%80%99s-math-problem/#

Whitebook, M., Sakai, L., Kipnis, F., Lee, Y., Bellm, D., Almaraz, M., & Tran, P. (2006). *California early care and education workforce study.* Center for the Study of Child Care Employment, University of California at Berkeley and California Child Care Resource and Referral Network. Retrieved from http://www.irle.berkeley.edu/cscce/wp-content/uploads/2006/01/statewide_highlights.pdf

2

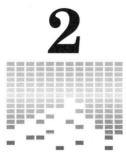

Practices, Knowledge, and Beliefs About Professional Development

Marilou Hyson and Taniesha A. Woods

In Chapter 1's framing-the-landscape discussion, Sharon Lynn Kagan and Rebecca Gomez described the general characteristics of the early childhood work force and set forth a daunting array of structural, conceptual, technological, and policy issues with which any discussion of professional development in early childhood mathematics must grapple. With this context in mind, in Chapter 2 we turn from the macro to the relatively micro, describing some of the everyday realities of professional development in early childhood mathematics.

Our Guiding Questions

1. What do teachers typically do in the classroom or other early childhood education (ECE) settings to promote young children's mathematical development?

2. What knowledge, beliefs, and attitudes might underlie their practices?

3. What mathematics-related professional development—preservice and inservice—is currently available to early childhood educators? Who is providing this professional development and with what content and methods?

4. What patterns emerge from exploring these realities that may prompt the future design and implementation of professional development in early childhood mathematics?

SETTING THE STAGE: TWO COMPOSITE SKETCHES

A recurring theme throughout this book is the lack of research on early childhood mathematics education (ECME). Compared with the number of studies of literacy education, for example, there is much less research in the domain of mathematics. This is an important omission because without understanding what currently exists, it is difficult to know where and how one might best intervene to encourage either better practices or wider implementation of the good practices that some teachers are using.

Although it is limited, the research we do have provides some sense of teachers' typical mathematics-related practices and the knowledge, beliefs, and professional preparation that may underlie these practices. From this research, we have created illustrative portraits of two typical early childhood practitioners—Maria and Adele:

Maria teaches 4-year-olds in a suburban child care center. When asked about mathematics in her classroom, she says math is in "everything that we do." She does not think that children at this age should be pushed toward formal math instruction. Much of the day is spent in free play. When the children are given materials to use such as blocks, construction toys, and puzzles, Maria has a sense that the children are picking up mathematical knowledge from using these materials. Of course, as in other full-day child care programs, the schedule includes a lot of time spent in toileting, meals, cleanup, and other routine tasks. Specific to math, Maria does encourage the children to count, and they practice counting every day at circle time and when lining up to go outside. Maria is starting on her associate's degree at the local community college, where she has taken several math courses—her "least favorite" subject and the one that she feels least comfortable with. She has not yet had a class in how to teach math to young children, although she would like to learn more.

Adele is a first-year kindergarten teacher in an urban public school in a low-income community. Every day she does math with her children using a curriculum that her district has recommended. She feels that there is a lot to cover, especially in her half-day kindergarten, but she tries her best to get through all the lessons. Most of the time she works on the curriculum with the whole class, and she uses the daily "calendar time" to teach math concepts. Adele also has a math table where children can choose different kinds of manipulative materials to use, mostly to help them learn counting and simple addition and subtraction. Recently, Adele has had quite a few immigrant children in her class; she feels that they are far behind in their math as well as literacy skills, and she is not sure that they have the ability to learn the concepts she is trying to teach them. Adele's bachelor of arts (B.A.) program did include several math courses and a math methods course, but it mainly focused on math for older elementary school children.

WHAT DO EARLY CHILDHOOD TEACHERS TYPICALLY DO TO PROMOTE CHILDREN'S MATHEMATICAL DEVELOPMENT?

With these composite portraits as a backdrop, we can turn to available data to get some sense of the amount of time early childhood teachers typically spend on math-related activities and instruction, the aspects of mathematics that they emphasize, and the methods that they use.

As highlighted in Chapter 1, the characteristics and diversity of the early childhood field make it especially difficult to generalize about these issues. In contrast to elementary and secondary education, practitioners in early childhood programs work in many different roles and settings. Practitioners may be known as teachers, child care providers, early interventionists, home visitors, family educators, and more. They may work in private or public preschools, child care centers, Head Start programs, family child care homes, state-funded prekindergarten programs, or kindergartens. These programs are under differing auspices and are subject to different sets of regulations, standards, and accountability systems. In part because of this, ECE practitioners are equally diverse in their level of formal education and in the nature of their professional preparation, as we saw when comparing Maria, who is just starting on her associate's degree, with Adele, who already has a B.A. in ECE.

Time Spent on Mathematics in Preschool and Kindergarten

Before looking specifically at math, it is important to underscore more general problems with the number and quality of learning opportunities in early childhood programs. As summarized in the National Research Council's (NRC) early childhood mathematics report (2009) and in ongoing work by Pianta and colleagues (2005), research has shown that prekindergarten and kindergarten children seldom receive adequate instructional support for their learning. Rather, children usually spend a great deal of time waiting in line, engaging in transitions from one activity to another, and participating in daily routines such as meals or hand washing. Although learning opportunities could be provided in any domain, including mathematics, during those times, the data showed that they were not.

On average, then, instructional support of any kind is lacking in typical early childhood programs. But within this context, it is mathematics that consistently takes a back seat. As detailed in a comprehensive literature review of the features of effective early childhood professional development interventions (U.S. Department of Education, 2010), "multiple studies noted that . . . most programs were providing extremely limited input in math" (p. 76).[1]

Because state-supported prekindergarten programs and public school kindergartens are subject to relatively stringent standards and accountability systems, one might assume that those programs offer more mathematics teaching than is found in settings such as child care centers and family child care homes. Few studies are available to verify that assumption, but there is evidence that state-funded pre-K programs may spend very little time on focused math activities, even when professional development has been provided to staff (Varol, Farran, Bilbrey, Vorhous, & Hofer, 2012).[2]

Content of Mathematics Activities:
Prekindergarten, Child Care, and Kindergarten

Unfortunately, we know little about the content of typical mathematics instruction in early childhood settings. In multistate prekindergarten observations conducted in a study by the National Center for Early Development and Learning (NCEDL), a wide array of activities were counted as "mathematics" (e.g., counting, talking about properties of shapes, sorting, measuring). However, as noted in the NRC (2009) report, the data did not include a breakdown of the primary emphasis or content focus within the broad category of mathematics.

Similar to the findings from this pre-K study, observations in community child care and family child care programs (Sarama & DiBiase, 2004) showed that teachers' main emphasis was on counting, sorting, numeral recognition, and patterning, with geometry and measurement being the least frequently mentioned topics. The description of Maria's child care program illustrates this emphasis. In kindergarten, most teachers—such as Adele—report that they try to cover *all* areas of math (sometimes even within 1 week), sacrificing depth for breadth and not always focusing on the competencies identified as being the most important foundations for later mathematical development (NRC, 2009).[3]

Methods Typically Used by Preschool Teachers

As with our knowledge of math content in preschool classrooms, much of our knowledge about teaching practices comes from observations done more than a decade ago as part of the multistate prekindergarten study conducted by NCEDL. As summarized in the NRC (2009) report, teachers were relatively unlikely to provide specific math-focused activities, relying on integration of math with other content (recall Maria's assertion that "math is in everything we do"). It is possible that focused math teaching has become more prevalent in preschool programs since the NCEDL study. In the intervening years, programs may have become more aware of the need for intentional, teacher-initiated math learning activities as part of developmentally appropriate practice (NRC, 2009; National Association for the Education of Young Children [NAEYC] and National Council of Teachers of Mathematics [NCTM], 2002).

Again drawing on research summarized in the NRC (2009) report, we also see that most math instruction—often about half of it—occurs in whole-group settings. Very little occurs in small groups, although research has underscored the benefits of small-group teaching in math and other areas (e.g., Clements & Sarama, 2008; Wasik, 2008). Whether in whole groups or small groups, prekindergarten teachers in the NCEDL study used a variety of teaching strategies. Some encouraged children's mathematical efforts, but many teachers also relied on didactic methods, asking questions with one correct answer or eliciting performance of isolated math skills. Instructional scaffolding was less often used than other teaching methods, despite evidence of its value in early childhood teaching (Bodrova & Leong, 2007; Copple & Bredekamp, 2009). Other research in pre-K programs serving low-income children confirms these observations (Frede, Jung, Barnett, Lamy, & Figueras, 2007). Again, there was little teacher support for math, even though these programs were rated as being of generally high quality. In addition, teachers in Frede and colleagues' study seldom used or encouraged the use of mathematics language in their classrooms, despite evidence that learning "math language" is especially important for low-income children (Jordan, Huttenlocher, & Levine, 1994; Levine, Suriyakham, Rowe, Huttenlocher, & Gunderson, 2010). And finally, providers of community-based child care and family child care services seldom include intentional teaching, instead relying on free play as a vehicle for mathematics learning.

Methods Typically Used by Kindergarten Teachers

Turning from methods used in preschool to those used by kindergarten teachers, the teacher surveys summarized in the NRC (2009) report offer some information. Again, the results are not encouraging. Kindergarten teachers said that they used many different instructional strategies, with verbal counting emphasized most frequently. Teachers often gave children manipulative materials to use when they practiced counting, but they seldom gave children these kinds of concrete tools to use for measuring. Kindergarten teachers (such as Adele) very often used daily calendar activities to teach math, yet most mathematics experts find that calendar activities are generally ineffective in teaching core mathematics concepts such as the base-10 system. An important reason is that the calendar is based on the 7 days of the week rather than the quantity 10, which underlies our number system (NRC, 2009).

The Use of Technology in Mathematics Teaching

Technology is a potentially valuable resource in many areas of early childhood teaching practice, including mathematics (Clements & Sarama, 2003; NAEYC & Fred Rogers Center, 2012). Several recent surveys show that only about one third of early childhood educators working with children younger than kindergarten age have digital media in their classrooms or family child care homes—far less than in K–12 settings. However, those early childhood teachers who have experience with digital media have a strong interest in using it (Grunwald Associates, 2009; Wartella, Schomburg, Lauricella, Robb, & Flynn, 2010). Like teachers of older children, early childhood educators primarily use digital media as a source of games and activities for children. Specific to mathematics, we know little about the extent to which early childhood educators are using digital media to support children's learning. One study of preschools in two states did indicate that few teachers—perhaps one out of three—used any kind of software in their mathematics teaching (Sarama & DiBiase, 2004). In the years since 2004, it is likely that the situation has changed somewhat, as technology itself has advanced along with teachers' exposure to its use in early education. Future research may document these changes and their impact on the field.

FROM CLASSROOM PRACTICES TO THE TEACHERS' MINDS: TEACHERS' MATH-RELATED KNOWLEDGE AND BELIEFS

With that brief, discouraging description of the mathematics-related time commitments, content emphases, and methods used by teachers in typical early childhood programs, we can now examine some factors that may underlie these patterns: teachers' knowledge of and beliefs about mathematics in ECE.

ECME is less familiar to many early childhood teachers than literacy, for example, yet we know that math is an essential part of children's current development and preparation for future academic success. When asked about mathematics education, many early childhood teachers report that they are uncomfortable with mathematics (Copley, 1999) and identify it as their weakest subject (Schram, Wilcox, Lapan, & Lanier, 1988). The provision of high-quality ECME requires that teachers become knowledgeable about the content, pedagogy, and curricular resources (NRC, 2009; Shulman, 1986). Furthermore, ECME necessitates that teachers understand young children's mathematical development—specifically, how it changes over time and how to align mathematics content, instruction, and curricular materials so that they provide children with sequential experiences that are challenging and achievable (NRC, 2009).

Current State of Early Childhood Teachers' Knowledge of Mathematics Education

The three areas of ECME that teachers must understand are 1) mathematics content knowledge, 2) mathematics pedagogical knowledge, and 3) children's mathematical development.

Mathematics content knowledge denotes concepts that a teacher must know as well as how and why these concepts are organized in the domain and how they relate to other concepts (Shulman, 1986). For example, understanding how to use numbers to quantify is fundamental to mathematics, but many adults, including early childhood teachers, are not aware of the complex thinking required to carry out enumeration or the idea of counting things—such as accurate knowledge of the number word list, the use of one-to-one correspondence, and understanding cardinality, competencies that are illustrated in detail in Chapter 3. A firm grasp of mathematical concepts ensures that teachers' implementation of math learning activities will be mathematically accurate.

Mathematics pedagogical content knowledge refers to knowledge of the subject matter and how to teach it as well as an understanding of what makes learning mathematics easy or difficult for students (Shulman, 1986). For example, to teach early number concepts, teachers who use a small number of concrete objects that are arranged in a straight line are more likely to have children who count with greater accuracy than teachers who put the objects in a pile. Placing the objects in a straight line facilitates young children correctly counting each object only once instead of multiple times, which is likely to happen when the objects are in a pile with no clear organization (NRC, 2009). Teachers need to develop realistic expectations for achievement based on knowledge of children's mathematical thinking and knowledge of how to support continued development by presenting more challenging mathematics (NRC, 2009; Lee, Hartman, Pappas, Chiong, & Ginsburg, 2011). In-depth discussions of children's mathematical development and teaching practices that support math learning are presented in Chapters 3 and 4, respectively.

To date, few empirical studies have investigated early childhood teachers' knowledge of mathematics education. Most of the research that directly measures knowledge of mathematics teaching and learning focuses on elementary and later grades (e.g., Ball, 1988, 1991, 2011; Ball, Sleep, Boerst, & Bass, 2009; Hill, Rowan, & Ball, 2005).

At the same time, there are a few studies that examine early childhood teachers' pedagogical content knowledge. In a recent study, McCray and Chen (2012) used the newly developed Pedagogical Content Knowledge for

Preschool Mathematics Interview to assess Head Start teachers' mathematics pedagogical content knowledge. McCray and Chen (2012) define pedagogical content knowledge as teachers' knowledge of mathematics content, teaching practice, and student development. To obtain an accurate sense of teachers' understanding, the interview uses realistic, classroom-based vignettes to assess whether teachers have the knowledge needed for effective teaching of foundational domains: number sense, patterns, operations, measurement, shape, spatial relationships, and classification (NCTM, 2000). For example, a vignette might show two children building an elephant pen with unit blocks. Based on what they observe, the teachers are asked to describe the math in the play, indicate what comments they could make to help the children see the math in their play, and note questions they could ask to help children find out more about the math in their play. Responding to the vignettes was challenging for all the Head Start teachers in McCray and Chen's study—even those who were high scoring.

However, there were variations in teachers' knowledge, with important correlates. Those teachers who showed greater pedagogical content knowledge on the interview were likely to use more math-related language in the classroom. Children in the classrooms of higher-scoring teachers also had better learning outcomes over the course of the year than children whose teachers had lower scores, suggesting that pedagogical content knowledge may be a key target for professional development.

To that end, another study provides some useful information about specific content areas in which teaches may benefit from additional support (Lee, 2010). Lee used the Survey of Pedagogical Content Knowledge in Early Childhood Mathematics to assess kindergarten teachers' knowledge in six domains: number sense, pattern, ordering, shapes, spatial sense, and comparison.[4] Teachers were most knowledgeable about number sense and pattern, followed by more limited knowledge about ordering, shapes, spatial sense, and comparison (Lee, 2010). The fact that number sense is the domain teachers were most knowledgeable about is not surprising. Recall our earlier example of Maria's classroom where her explicit math instruction focused on having the children count, an emphasis that is consistent with research that shows number (together with simple addition and subtraction) is most frequently taught in early childhood classrooms (NRC, 2009).

The Committee on Early Childhood Mathematics (NRC, 2009) and NCTM's Curriculum Focal Points (2006) state that mathematics education should reflect a wider breadth of domains, including number and operations (addition/subtraction situations), geometry (shape and spatial relations), measurement, and pattern—all in the context of children solving problems and communicating their mathematical thinking. Yet the reality is that teachers teach what they know, and their knowledge is often narrow.

Lee (2010) did find that teachers with advanced degrees and with more extensive kindergarten teaching experience showed more pedagogical content knowledge than those with lower levels of education and experience. As noted earlier in this chapter, few teachers at any experience level engage in any actual mathematics teaching, yet it does seem that the need for ongoing professional development is especially acute for less experienced, less educated staff.

An additional area of research and practice that has received relatively little attention is teachers' knowledge about children's mathematical development. Few empirical studies examine what early childhood teachers know about children's developmental trajectories as they apply to math (see Clements & Sarama, 2009 for more information on math learning trajectories). In fact, the NRC (2009) report is one of the few publications to make this a primary aim of its work: providing detailed guidance on children's developmental trajectories or "teaching–learning paths" (p. 2) for mathematics. Recent research reinforces the need for early childhood practitioners to receive training and education on mathematical development. Woods (2011) found that teachers of 3- and 4-year-olds report the need for more support in the organization and sequence of their math curriculum so that it aligns with children's mathematical development and includes the important math topics necessary for success in kindergarten.

Early Childhood Teachers' Beliefs and Attitudes Toward Mathematics Education

In addition to examining teachers' knowledge of ECME, attention should be paid to their attitudes toward and beliefs about mathematics teaching and learning for young children. Early childhood educators' beliefs are associated with their teaching practices (Charlesworth, Hart, Burts, & Hernandez 1991; Charlesworth et al., 1993; Pianta et al., 2005; Stipek & Byler, 1997; Stipek, Givvin, Salmon, & MacGyvers, 2001), and teachers' beliefs serve as the filter through which they process classroom interactions and shape their instruction (NRC, 2009). Historically, U.S. teachers believed that mathematics was a static body of knowledge relying on manipulating rules and procedures (Thompson, 1992). For many early childhood teachers, this belief still influences their mathematics instructional practice. Across a series of studies, findings indicate that early childhood teachers believe social and emotional development, physical well-being, and literacy are more important than mathematics learning when it comes to foundational skills (Ginsburg et al., 2006; Ginsburg, Lee, & Boyd, 2008; Lin, Lawrence, & Gorell, 2003; Piotrkowski, Botsko, & Matthews, 2001). Teachers generally believe that math learning should be embedded in everyday routines so that children have opportunities to learn it throughout the day (Lee & Ginsburg, 2007). However, differences in teachers' beliefs about math instruction appear when students' socioeconomic backgrounds are considered—we return to this issue later in the chapter.

Part of the reason teachers do not provide rich and varied mathematics experiences for young children may be related to the commonly held belief that ECME is developmentally inappropriate (NRC, 2009), a belief that many educators held in the past and that still persists today. This belief may also be related to practitioners' lack of familiarity with high-quality mathematics instruction. Some seem to equate math teaching with rote learning and drills rather than understanding that these rigid instructional methods do not, and need not, define early mathematics education. Instead, "academic" material can be taught in developmentally appropriate ways; to this end, as recommended in the NRC (2009) report, early childhood educators, including professional development specialists, need education and training on how to use intentional instruction to promote high-quality mathematics teaching and learning.

In math, as in other areas, beliefs do seem to influence behavior. Lee (2005) found that kindergarten teachers with positive attitudes toward teaching mathematics were likely to report the use of developmentally appropriate mathematics instruction that was child centered, included authentic assessment, and used appropriate manipulatives (Lee, 2005). By contrast, Trice and Ogden (1987) found that elementary school teachers with anxiety and negative attitudes toward teaching mathematics spent less time on mathematics instruction. In another study, early elementary teachers, who are predominantly female, held beliefs about math that were related to their female students' beliefs about their own competence and their achievement scores in math (Beilock, Gunderson, Ramirez, & Levine, 2010). That is, girls in the classrooms of math anxious female teachers were more likely to believe that boys are good in math and girls are good in reading. Moreover, these girls had lower math achievement scores; the same results were not found for boys. The findings from these studies have implications for what we see in early childhood classrooms: If early childhood teachers are uncomfortable with mathematics and hold negative attitudes toward it—like Maria in our earlier example—they will be less likely to spend instructional time on it. And if these negative beliefs about math are reinforced and expanded in the early elementary grades, they are especially problematic for young girls. Overall, the findings from these studies suggest that teachers' beliefs about and attitudes toward math influence their practice and have the potential to shape children's beliefs about and achievement in math.

The good news is that negative attitudes are subject to change: Evidence suggests that improvement in teachers' math content knowledge is positively linked to their attitudes and beliefs toward teaching and their own learning of math (Ben-Chaim, Keret, & Ilany, 2007). Taken as a whole, these findings suggest that teacher education and training should include an examination of beliefs about and attitudes toward mathematics education and intentional efforts to shift those beliefs and attitudes in a more positive direction.

Teachers' Beliefs About Mathematics Education in the Context of Sociodemographic Diversity

In addition to research on teachers' beliefs about mathematics education, a small body of research has also examined how teachers' beliefs and expectations about children's mathematical competence and about mathematics instruction are related to factors such as children's socioeconomic status (SES) and race/ethnicity. This body of work is important because teachers who expect their students to be successful are more likely to have students that confirm these expectations. Conversely, teachers who do not believe in their students' mathematical competence or intellectual ability because of children's race/ethnicity, socioeconomic, or language status—like our composite teacher Adele—may have their low expectations confirmed through student performance and/or how it is assessed (Ladson-Billings, 1997).

Lee and Ginsburg (2007) found that teachers of low-SES children of color (almost entirely African American and Hispanic) believe that mathematics instruction is critical to preparation for kindergarten and should be implemented even if young children initially show little interest. Conversely, teachers of middle-SES children (almost entirely white) believe that mathematics instruction should be child initiated, child centered, and encouraging of children's socioemotional development (Lee, 2006; Lee & Ginsburg, 2007). The difference in teachers' beliefs about mathematics instruction for different groups of children may be related to their perceptions of parental practices around education and the K–12 educational system. That is, teachers of middle-SES children may perceive significant pressure being put on children at home to excel academically, so they do not want to pressure children in the classroom (Lee & Ginsburg, 2007). Furthermore, teachers of low-SES children may perceive that schools serving these students are ill-prepared to meet children's individual needs for a variety of reasons, including fewer resources than those enjoyed by middle-SES schools. These teachers may also be aware of the persistent trends showing that low-SES children who enter school behind their middle-SES peers typically stay behind (Reardon, 2011).

One interpretation of these findings is that both sets of beliefs reflect teachers' perception about some realities of American education. As a group, low-SES children, especially children of color, perform more poorly in school than do middle-SES children, and their public school education is likely to be of lower quality than that received by middle-SES children. Given these facts, teachers may conclude that low-SES children of color need more intensive preschool instruction than do middle-SES children. Similarly, if teachers perceive that the parents of middle-SES children exert excessive pressure on their children to learn, then their teachers may feel that those children could benefit from a more laissez-faire approach in school.

Whatever the merits of these interpretations, two things are clear: First, issues of race and social class do influence teachers' beliefs about children's math education, and second, teachers of both lower- and middle-SES children can benefit from a deeper understanding of children's mathematical competence and the nature of developmentally appropriate math education.

PREPARING EARLY CHILDHOOD EDUCATORS: EXISTING PROFESSIONAL DEVELOPMENT OPPORTUNITIES

The mathematics-related classroom practices, knowledge, and beliefs of typical early childhood teachers like Maria and Adele are surely influenced by the pre- and in-service professional development opportunities that they have received—or, often, not received. In this section of the chapter, we describe those opportunities as they presently exist; Chapters 6 and 7 discuss promising practices and recommendations for the future of pre- and in-service professional development. We begin with a look at the status of mathematics education in higher education programs and then turn to ongoing or in-service professional development.

HIGHER EDUCATION AND THE PREPARATION OF FUTURE TEACHERS OF EARLY CHILDHOOD MATHEMATICS

High-quality preservice teacher preparation in mathematics was noted as an "urgent priority" in *Early Childhood Mathematics: Promoting Good Beginnings,* a joint position statement of the National Association for the Education of Young Children (NAEYC) and the National Council of Teachers of

Mathematics (NCTM; 2002, p. 10) as well as in an editorial in the *Journal of Early Childhood Teacher Education* (Benner & Hatch, 2009).

Mathematics Content in Higher Education Programs

In their joint position statement, NAEYC and NCTM recommended the following:

> To support children's mathematical proficiency, every early childhood teacher's professional preparation should include these connected components: (1) knowledge of the mathematical content and concepts most relevant for young children—including in-depth understanding of what children are learning now and how today's learning points toward the horizons of later learning; (2) knowledge of young children's learning and development in all areas—including but not limited to cognitive development—and knowledge of the issues and topics that may engage children at different points in their development; (3) knowledge of effective ways of teaching mathematics to all young learners; (4) knowledge and skill in observing and documenting young children's mathematical activities and understanding; and (5) knowledge of resources and tools that promote mathematical competence and enjoyment. (NAEYC & NCTM, 2002, p. 10)

We have little information about the extent to which these components are included in most early childhood teacher educational programs (Horm, Hyson, & Winton, 2013; Hyson, 2008), but what we do know should be a source of concern. Responses to several surveys of faculty and other leaders in higher education programs provide insight about the extent to which mathematics content is emphasized in early childhood teacher preparation (e.g., Early & Winton, 2001; Hyson, 2008; Johnson, Fiene, McKinnon, & Babu, 2010; Maxwell, Lim, & Early, 2006). A note of caution: These surveys were conducted for various purposes and with different higher education samples, only some of which were representative. In addition, we do not know how accurate survey participants' responses were about their own programs and practices; people understandably may want to put things in a positive light. With these limitations in mind, the survey results, as a whole, show some consistent patterns.

Lack of Robust Course Requirements in Mathematics Education

To a substantial extent, faculty members reported that both state early learning guidelines and teacher certification or licensure requirements influence decisions about math course offerings (Hyson, 2008). Although it is difficult to determine specific state-level licensure requirements (and many of these requirements are being replaced by state approval of teacher educational programs or by state competency tests for graduating teachers), state requirements give more attention to required coursework in mathematics content than to required courses in teaching methods. A majority of faculty members report that these kinds of requirements have a strong influence on their program offerings (Hyson, 2008). Possibly as a result, virtually all B.A. programs require specific mathematics content courses for future teachers, as do most associate's degree programs, while giving less attention to courses on mathematics education, including how to teach math in early childhood settings.

Associate's and bachelor's degree programs use a range of approaches to mathematics education for early childhood students (Hyson, 2008): Some require one or more courses in teaching early childhood mathematics, whereas others embed mathematics education in general early childhood curriculum coursework or combine mathematics and science education. Furthermore, some institutions of higher education focus only on mathematics education for elementary school teaching, as Adele's B.A. program did.

In sum, preservice teachers do not have access to robust course offerings in ECME. Lacking this background, it is not surprising that even degreed teachers like Adele either de-emphasize mathematics instruction or teach math in ways that are inconsistent with current knowledge.

In addition, although mathematics and literacy should be viewed as complementary rather than competitive domains (NRC, 2009), most institutions require fewer methods courses in mathematics than in literacy.[5] Even if preservice programs include a math methods course, they are likely to put more emphasis on helping students define their teaching philosophy and learn some instructional methodologies rather than providing specific focus on the mathematics content or children's mathematical development (Ginsburg et al., 2006).

Looking at these patterns, it is tempting to conclude that future early childhood teachers should be required to take more courses in math and math methods. This kind of remedy was often applied by states in response to concerns about teachers' lack of knowledge and skill in the domain of early literacy. However, data from the field of K–12 math education indicate that simply requiring more courses and credit hours does little good. What is needed instead is to ensure that course content and pedagogy are designed to help future teachers implement practices likely to produce positive math outcomes for children (Conference Board of the Mathematical Sciences, 2001)—points that are richly illustrated in Chapter 6.

Limited Opportunities for Math-Related Field Experiences

In its 2010 report *Preparing Teachers: Building Evidence for Sound Policy,* the NRC described field experiences (or "clinical preparation") as one of the components of teacher education that has the greatest likelihood of improving child outcomes. Similar points have been made by the National Council for Accreditation of Teacher Education (NCATE), for many years the leading organization providing accreditation for teacher-preparation programs (2010). However, solid descriptive information is not available on hours and settings used for early childhood field experiences, including student teaching. Reflecting the more general concerns in NCATE's report, a survey and interview study of higher education programs for early educators in 40 states found that relatively few programs aligned content topics directly with clinical practice (Johnson et al., 2010).

Turning specifically to mathematics, across these sources of data we also see that relatively few programs link mathematics coursework to direct experiences with children. Even programs applying for NCATE and NAEYC accreditation have been unlikely to include any mathematics-specific field experience, although students might be required to implement some math lessons during student teaching (Hyson, 2008).

Faculty Perceptions: Students Are Underprepared and Faculty Members Need Help

Several higher education surveys suggest that early childhood faculty members are not satisfied with their programs' ability to prepare students in math. Hyson (2008) found that about 40% of B.A. program faculty and 58% of associate's degree faculty thought their graduates were only "somewhat prepared" to promote children's mathematical skills and interest. Johnson and colleagues (2010) asked faculty to rank the quality of their programs' preparation of future teachers across different content areas. Overall, faculty members rated their own program highest in the domain of literacy, with math holding sixth place.

In Hyson's (2008) survey, faculty did not propose additional mathematics-related courses as a solution, although more than half wanted better field experience opportunities for their students, and many wished that their students had better preparation in mathematics before they entered higher education.

Faculty members' answers to some survey questions indicated gaps in their own knowledge, for example, about mathematics curriculum and about essential competencies for future early childhood teachers. Fortunately, many faculty members indicated that they want to know more about mathematics and are eager to learn how better to prepare their students to teach math to young children (Hyson, 2008). Some also emphasized that faculty development must bridge institutional divides between specialized math faculty and ECE faculty, as well as between full-time and adjunct instructors. As one faculty member said, "[We need] math faculty who understand the needs of early childhood and can do more than just teach the subject . . . without further turning the students off to the subject of math . . . too often we only reinforce pre-service teachers' dislike/fear of math!" (p. 37).

Faculty members, especially those in associate's degree programs, are also eager to obtain resources to help them prepare students to teach math. For example, one respondent said that videos showing children engaged in mathematics (and enjoying it) are the most valuable tools in her undergraduate teaching. With those needs in mind, Chapter 6 includes a number of examples of the innovative use of video-based instruction in preservice education—resources that would surely help enrolled college students like Maria and Adele gain greater enthusiasm for math and greater skill in giving children engaging, effective learning opportunities.

In-Service Professional Development in Mathematics: A Barren Landscape

Even the best preservice preparation is unlikely to result in effective teaching without ongoing professional development. Chapter 5 outlines the features of effective professional development. Among other things, it must be ongoing and, to a great extent, embedded in the job (U.S. Department of Education, 2010). Yet the two teachers described at the beginning of this chapter are typical of the early childhood work force in their limited opportunities to gain essential math-related knowledge and skills through in-service professional development. No wonder kindergarten teacher Adele has few resources to help her enrich and deepen her children's math experiences, and that child care teacher Maria has difficulty going beyond counting activities.

As described in Chapter 1, the early childhood system is diverse and complex, with services for young children delivered in multiple settings, each with its own auspices and regulations. As Kagan and Gomez note in Chapter 1, most states require some amount of in-service professional development for teachers. However, the specifics vary greatly, and there is little evidence of specific math-related expectations for ongoing professional development. What follows are brief descriptions of what may be required and available for early childhood staff who work in different kinds of settings.

Public School Kindergarten–Primary Programs

Within the public school system, states and school districts set their own expectations for in-service professional development. Beyond the preservice requirement of a bachelor's degree and teacher certification, most states require annual professional development. Many states set standards for professional development activities, and some provide funds to school districts for in-service professional development. Little or no information is available about what is usually required in mathematics, but as in other content areas, the overall professional development pattern is one of isolated workshops given by "math experts," without sufficient opportunities to observe and implement specific teaching practices. Our description of kindergarten teacher Adele having received no math in-service training is probably typical. With the implementation of the K–12 Common Core State Standards in mathematics (National Governors Association Center for Best Practices [NGA] & Council of Chief State School Officers [CCSSO], 2010), it is likely that more professional development will be provided to kindergarten teachers in this content area.

State Prekindergarten Programs

According to Brenneman, Stevenson-Boyd, and Frede (2009), among the 50 state-funded preschool programs included in their report, 41 require at least 15 hours of in-service training per year. However, decisions regarding content priorities or requirements are usually made locally. Teachers working in state pre-K programs, then, may or may not have opportunities for professional development in mathematics.

Head Start Programs

As described in the Improving Head Start for School Readiness Act of 2007 (Head Start Act of 2007; PL 110-134), all teachers must have regular in-service training:

> Each Head Start teacher shall attend not less than 15 clock hours of professional development per year. Such professional development shall be high-quality, sustained, intensive, and classroom-focused in order to have a positive and lasting impact on classroom instruction and the teacher's performance in the classroom, and regularly evaluated by the program for effectiveness.

Beyond this general statement, the directive does not specify the content of in-service professional development. Head Start's Child Development Outcomes include expectations for children's mathematical development; however, these are not explicitly linked to expectations for staff training. The Office of Head Start has produced a series of webcasts to help supervisors, directors, and technical assistance providers learn more about the Head Start mathematics outcomes and how to support staff in promoting those outcomes. In addition, through Head Start's relatively new Early Childhood Learning and Knowledge Center (ECLKC), recommendations have been disseminated about effective models of focused professional development in mathematics (Copley, 2007).

Child Care Centers that Are Not Part of State Pre-K or Head Start Programs

In Chapter 1, Kagan and Gomez note that even in those states that require a certain number of annual professional development hours for child care center staff, there are no mathematics-specific requirements. Although professional development has usually been provided in a series of disconnected, one-time workshops (what some have described as "drive-by training" or "spray and pray"), some states are moving toward more in-depth professional development connected to early learning guidelines (ELGs), carrying at least the potential for greater mathematics education content, although states' attention to mathematics in their ELGs varies dramatically (Scott-Little, Kagan, Reid, & Castillo, 2011).

Family Child Care Homes

Those who work in state-licensed family child care homes often (but not always) must meet requirements for annual hours of professional development, although the requirements are even fewer than for nonstate-licensed child care center staff. At present, there are no expectations that family child care providers should have any mathematics-specific training. A recent survey by the National Association of Child Care Resource and Referral Agencies (NACCRRA; 2012) showed that 19 states did require that some of the annual family child care training hours must be in the category of "learning activities," creating at least the potential for professional development in mathematics education.

Looking Across Settings: In-Service Professional Development and State Quality Rating and Improvement Systems

As described in Chapter 1, since the 1990s, many states have established Quality Rating and Improvement Systems (QRIS) in an effort to assess, improve, and communicate to the public the level of quality in early childhood services within the state. As of February 2014, 37 states and the District of Columbia had launched statewide QRISs (QRIS National Learning Network, 2014). Stars, or rating categories, define levels of quality, and one aspect of the quality rating has to do with the extent to which staff working in a specific ECE program have achieved certain levels of education and ongoing professional development.

In a compendium of characteristics of state QRISs, Tout and colleagues (2010) summarized information on QRIS-related professional development expectations. Of the 18 states that provided a description of the content of this professional development, none specified mathematics. The most commonly reported professional content had to do with assessment of the early childhood program environment. This was followed by language and literacy, specific curriculum, business practices, safety, and social and emotional development. Furthermore, although a number of states provide "quality improvement activities" to help programs attain higher ratings, most of these are not specifically targeted to any content areas or specific teaching practices (Tout, 2013).

It is worth mentioning that state QRISs are increasingly aligning their professional development expectations with their state ELGs, all of which include mathematics competencies in their child outcomes. In addition, the K–12 Common Core State Standards in the standards categories of Mathematics and English Language Arts (NGA & CCSSO, 2010) are beginning to exert considerable influence below kindergarten, as states align or revise their existing early learning guidelines in light of the Common Core. This may, in the future, be a mechanism through which the Common Core can indirectly affect what is emphasized in in-service ECME professional development.

CONCLUSION: FROM EVERYDAY REALITIES TO REALIZING POTENTIALS

The goal that motivates this book is to ensure that all young children experience learning opportunities that support their mathematical development—including foundational knowledge, skills, and interest and engagement in mathematical thinking. For this to happen, early childhood educators with positive attitudes toward math need to implement practices that are continuously informed by children's development, relying on intentional teaching and formative assessment and including achievable yet challenging activities that promote important learning objectives.

However, this chapter has illustrated how the everyday realities of ECE are, for the most part, getting in the way of positive math outcomes rather than promoting them. Beginning with the composite descriptions of teachers Maria and Adele, we have seen that research—although limited—is consistent in the picture it portrays:

- Very little attention is paid to mathematics in early childhood programs. This is the case across different kinds of settings and auspices, including those in which teachers have relatively high levels of formal education.

- In the math teaching that does occur, the content is not aligned with what is recommended (NRC, 2009). For example, preschool teachers emphasize rote counting and numeral recognition while often neglecting other key areas. By the time children reach kindergarten, there may be more breadth of coverage but at the expense of depth and focus. In addition, activities such as "calendar time," although perceived as mathcentric, lack real mathematical benefits.

- Methods used to teach math often rely on either whole-group teaching or incidental math experiences integrated into other activities. Small groups and instructional scaffolding are underused given their demonstrated effectiveness, and teachers seldom promote the use of math language in the classroom.

- Early childhood teachers, on average, know little about math content or how best to teach math to young children. They also lack knowledge of how children's mathematical thinking develops.

- Most teachers dislike math, are afraid of it, and do not like to teach it. They may believe it is less important than other aspects of the early childhood curriculum or that intentional math instruction is not developmentally appropriate. Teachers' beliefs about what math education should include also vary depending on the sociodemographic characteristics of the children in their classrooms.

- Professional development could potentially help educators gain important knowledge, skills, and positive attitudes toward math. However, neither preservice nor in-service professional development is currently up to this task.

- In higher education, early childhood programs lack strong mathematics course requirements, and existing courses often stress theory over practice or fail to align course content with young children's mathematical development. Field experiences focused on math education are lacking. Faculty themselves admit that their students are not as well-prepared in math education than in other areas; many faculty also acknowledge a need to strengthen their own knowledge base in ECME.

- Because early childhood programs are under so many different auspices, in-service requirements vary tremendously. However, math is almost never a required element in ongoing professional development.

Given this limited access to high-quality, evidence-based professional development, it is not surprising that teachers like Maria and Adele do not provide engaging, worthwhile math experiences for the children in their classrooms. We know that professional development *can* be effective in producing improved teaching practices and better outcomes for young children (U.S. Department of Education, 2010; Zaslow, Chapter 5). However, we lack broadly accessible, well-evaluated, professional development opportunities for all early childhood educators—professional development that creates not only knowledge of math and math teaching but also genuine joy and curiosity about mathematics—dispositions that will then be communicated to the children they teach.

Other chapters in this book provide examples of this kind of professional development. Our assessment of the everyday realities suggests that certain issues should be given high priority.

First, mathematics education should be infused into all early childhood professional development, not as an occasional option but as an integral part of what all early childhood educators should know and be able to teach. The primary focus of that professional development should be on promoting positive mathematics-related outcomes for all children. Second, much more extensive, well-supervised field experiences or on-the-job coaching sessions are needed so that teachers can practice new skills and receive supportive feedback. Third, professional development will need to help teachers implement curriculum that emphasizes the knowledge and skills most important for children's progress, keeping developmental trajectories in mind as well as children's individual characteristics and learning needs. Fourth, attention is needed to building the capacity of professional development providers, for both college and university faculty as well as community-based trainers, so that they are able to implement content and methods that can improve the knowledge, skills, and attitudes of future and current early childhood educators.

Although this chapter does paint a discouraging picture, there are reasons for optimism. After decades of an almost exclusive focus on literacy, mathematics and the preparation of teachers in the mathematics domain are receiving much greater attention. Potential levers for change include the federal government's focus on better access to high-quality ECE programs for low- and middle-income families and the renewed focus on science, technology, engineering, and math education; the K–12 Common Core State Standards in mathematics (adopted by 45 states and the District of Columbia); recommendations for reforms in the content and assessment of teacher education, moving toward a much stronger clinical or practice orientation (e.g., NCATE, 2010); and states' use of their QRISs to improve the content and quality of in-service professional development. But perhaps the greatest source of optimism is the early childhood work force itself. Often overworked and underpaid, educators like Maria and Adele remain deeply committed to their children and sincerely interested in practical resources to help with their work—in teaching mathematics as well as other areas of early development and learning.

FOR REFLECTION AND ACTION

1. The chapter begins with composite sketches of two teachers in ECE programs. To what extent do the descriptions of Maria's and Adele's practices, attitudes, and professional development opportunities match your impressions or experiences? (If you are not currently in the classroom, you may want to visit some programs and interview some teachers.)

2. In your experience (or those with whom you discuss these issues), what are the main obstacles to implementing effective practices in teaching math to young children?

3. The authors of this chapter share evidence that there is a need to strengthen the knowledge and skills of those who actually provide mathematics-related professional development—for college and university faculty as well as those who implement early childhood professional development within organizations or communities. Again, is that research evidence consistent with your experience?

4. Consider how to gather information about existing professional development opportunities for current or future ECE teachers within your state, district, or community. In doing so, you might reach out to colleagues to help collect this information and then to discuss its implications for future planning.

5. If you are a policy maker or other education decision maker, it will be especially important to identify the future professional development needs and priorities within your state, district, or community. Using or adapting the "Early Childhood Math Education Action Planning Tool" found in the downloadable content, begin planning how to

conduct this kind of needs assessment. Note that other chapters help you dig deeper into these issues.

CHAPTER 2 NOTES

1. Our information was based on observational data from prekindergarten and kindergarten classrooms that were collected by the National Center for Early Development and Learning (NCEDL) and from teacher surveys from the Early Childhood Longitudinal Study-Kindergarten (ECLS-K), both summarized in the National Research Council (NRC) report (2009) and several smaller studies. Despite variations in sample size, data sources, and metrics used, the message is consistent: Math receives extremely short shrift in U.S. early childhood programs.
   In the NCEDL study, mathematics of any sort occupied only 6% of the prekindergarten day, and in kindergarten, it was still only 11%. In comparison, children spent 14% of their time on literacy in pre-K and 28% in kindergarten—still low, but substantially more than in mathematics. Another time estimate at the kindergarten level comes from a teacher survey conducted as part of the ECLS-K study, again summarized in the NRC report. Most teachers (81%) said math is part of their daily practice, whereas 65% said that math occupies at least 30 minutes per day.
2. These researchers used observational data from a small number of urban pre-K and Head Start classrooms to determine the number of intentional, math-specific activities—anything more than 30 seconds long was counted—led by teachers each day, comparing a randomly assigned treatment group receiving professional development on a math curriculum with a control group. Both groups conducted only one or two math-related activities in a typical day, with some control group teachers providing no math activities at all across the 3 days of observation.
3. The survey of kindergarten teachers conducted as part of the ECLS-K study provides more information on the question of content emphases, although with the usual cautions about self-report data. From a list of 27 math concepts and skills, teachers reported providing instruction on *all* the areas included in the National Council of Teachers of Mathematics (NCTM) standards during a typical week (although one does wonder if teachers reported what they thought researchers wanted to hear)—in contrast to professional recommendations about providing greater depth and focus in mathematics teaching (NRC, 2009; NGA & CCSSO, 2010). Relatively more attention was given to number and operations than any other area, although the specific content emphases within that area did not, according to the authors of the NRC report, reflect the recommended balance. Finally, the domain of geometry received quite limited attention, despite its being identified as a focus area both by NCTM and now in the Common Core State Standards for kindergarten (NGA & CCSSO, 2010).
4. Lee (2010) states that the Survey of Pedagogical Content Knowledge in Early Childhood Mathematics was primarily selected for this study because of a lack of instruments that measure early childhood teachers' mathematical pedagogical content knowledge. The instrument is research based and comes from Smith's unpublished manuscript (1998) and dissertation (2000).
5. For example, Maxwell and colleagues (2006) found disparities in both associate's programs (65% required at least one literacy course versus 49% for math) and B.A. programs (77% literacy versus 59% math). Johnson and colleagues' survey of 40 B.A. ECE programs in research institutions showed that 38 programs (95%) required coursework in literacy methods but that only 29 (72%) had similar requirements in math methods. Johnson and colleagues also found that math methods courses more frequently included a wider span of development than did the corresponding literacy methods offerings, which often provided separate courses addressing children below kindergarten age. Hyson's survey (2008) showed similar results with a larger and more diverse sample; furthermore, those results showed that very few B.A. programs (6%) required more than one math methods course, although in the field of literacy, multiple methods courses were frequently required.

REFERENCES

Ball, D.L. (1988). *Knowledge and reasoning in mathematical pedagogy: Examining what prospective teachers bring to teacher education.* (Unpublished doctoral dissertation). Michigan State University.

Ball, D.L. (1991). Research on teaching: Making subject matter knowledge part of the equation. In J.E. Brophy (Ed.), *Advances in research on teaching* (Vol. 2, pp. 1–48). Greenwich, CT: JAI Press.

Ball, D.L. (2011). *Knowing mathematics well enough to teach it: Mathematical Knowledge for Teaching (MKT).* Paper presented at the Israel Academy of Sciences and Humanities, Jerusalem, Israel.

Ball, D.L., Sleep, L., Boerst, T.A., & Bass, H. (2009). Combining the development of practice and the practice of development in teacher education. *Elementary School Journal, 109*(5), 458–474.

Beilock, S.L., Gunderson, E.A., Ramirez, G., & Levine, S.C. (2010). Female teachers' math anxiety affects girls' math achievement. *Proceedings of the National Academy of Sciences, 107*(5), 1860–1863.

Ben-Chaim, D., Keret, Y., & Ilany, B. (2007). Designing and implementing authentic investigative proportional reasoning tasks: The impact on pre-service mathematics teachers' content and pedagogical knowledge and attitudes. *Journal of Mathematics Teacher Education, 10,* 333–340.

Benner, S.M., & Hatch, J.A. (2009). From the editors: Math achievement and early childhood teacher preparation. *Journal of Early Childhood Teacher Education, 30*(4), 307–309.

Bodrova, E., & Leong, D.J. (2007). *Tools of the mind: The Vygotskian approach to early childhood education* (2nd ed.). Upper Saddle River, NJ: Pearson Education/Merrill.

Brenneman, K., Stevenson-Boyd, J.S., & Frede, E. (2009, March). *Math and science in preschool: Policies and practice* (Preschool Policy Brief, Issue No. 19). New Brunswick, NJ: National Institute for Early Education Research. Retrieved from http://nieer.org/resources/policybriefs/20.pdf

Charlesworth, R., Hart, C.H., Burts, D., & Hernandez, S. (1991). Kindergarten teachers' beliefs and practices. *Early Development and Care, 70*(1), 17–35.

Charlesworth, R., Hart, C.H., Burts, D., Thomasson, R.H., Mosley, J., & Fleege, P.O. (1993). Measuring the developmental appropriateness of kindergarten teachers' beliefs and practices. *Early Childhood Research Quarterly, 8*(3), 255–276.

Clements, D.H., & Sarama, J. (2003). Young children and technology: What does the research say? *Young Children, 58*(6), 34–40.

Clements, D.H., & Sarama, J. (2008). Experimental evaluation of the effects of a research-based preschool mathematics curriculum. *American Educational Research Journal, 45*(2), 443–494.

Clements, D.H., & Sarama, J. (2009). *Learning and teaching early math: The learning trajectories approach.* New York, NY: Routledge.

Conference Board of the Mathematical Sciences. (2001). *The mathematical education of teachers.* Providence, RI: American Mathematical Society and Mathematical Association of America.

Copley, J.V. (Ed.). (1999). *Mathematics in the early years.* Reston, VA: National Council of Teachers of Mathematics.

Copley, J. (2007). *Focused professional development for teachers of mathematics.* (Professional Development, Head Start Bulletin No. 79.) HHS/ACF/OHS. Retrieved from http://eclkc.ohs.acf.hhs.gov/hslc/hs/resources/pd/Organizational%20Development/Training%20and%20Technical%20Assistance/FocusedProfessio.htm

Copple, C., & Bredekamp, S. (Eds.). (2009). *Developmentally appropriate practice in early childhood programs serving children from birth through age 8* (3rd ed.). Washington, DC: National Association for the Education of Young Children.

Early, D.M., & Winton, P.J. (2001). Preparing the workforce: Early childhood teacher preparation at 2- and 4-year institutions of higher education. *Early Childhood Research Quarterly, 16*(3), 285–306.

Frede, E., Jung, K., Barnett, W.S., Lamy, C.E., & Figueras, A. (2007). *The Abbott Preschool Program Longitudinal Effects Study (APPLES)*. Rutgers, NJ: National Institute for Early Education Research.

Ginsburg, H.P., Kaplan, R.G., Cannon, J., Cordero, M.I., Eisenband, J.G., Galanter, M., & Morgenlander, M. (2006). Preparing Early Childhood Educators to Teach Mathematics. In M. Zaslow & I. Martinez-Beck (Eds.), *Critical issues in early childhood professional development* (pp. 171–202). Baltimore, MD: Paul H. Brookes Publishing Co.

Ginsburg, H.P., Lee, J., & Boyd, J.S. (2008). Mathematics education for young children: What it is and how to promote it. *Society for Research in Child Development Social Policy Report, 22*(1). Retrieved from http://www.srcd.org/sites/default/files/documents/22-1_early_childhood_math.pdf

Grunwald Associates. (2009). *Digitally inclined*. Annual survey of educators' use of media and technology. Arlington, VA: Public Broadcasting Service.

Hill, H.C., Rowan, B., & Ball, D.L. (2005). Effects of teachers' mathematical knowledge for teaching on student achievement. *American Educational Research Journal, 42*(2), 371–406.

Horm, D.M., Hyson, M., & Winton, P.J. (2013). Research on early childhood teacher education: Evidence from three domains and recommendations for moving forward. *Journal of Early Childhood Teacher Education, 34*(1), 95–112.

Hyson, M. (2008). *Preparing teachers to promote young children's mathematical competence*. Paper commissioned by the Committee on Early Childhood Mathematics. Washington, DC: National Research Council.

Improving Head Start for School Readiness Act of 2007, 42 U.S.C. §§ 9801 *et. seq.* Sec. 648A Staff Qualifications and Development. Retrieved from http://eclkc.ohs.acf.hhs.gov/hslc/tta-system/ehsnrc/Early%20Head%20Start/supervision/staffing/Sec648AStaff.htm

Johnson, J., Fiene, R., McKinnon, K., & Babu, S. (2010). *A study of ECE pre-service teacher education at major universities in 38 preK states*. Final report to the Foundation for Child Development. State College: Pennsylvania State University.

Jordan, N.C., Huttenlocher, J., & Levine, S.C. (1994). Assessing early arithmetic abilities: Effects of verbal and nonverbal response types on the calculation performance of middle- and low-income children. *Learning and Individual Differences, 6*(4), 413–432.

Ladson-Billings, G. (1997). It doesn't add up: African American students' mathematics achievement. *Journal for Research in Mathematics Education, 28*(6), 697–708.

Lee, J. (2005). Correlations between kindergarten teachers' attitudes toward mathematics and teaching practice. *Journal of Early Childhood Teacher Education, 25*(2), 173–184.

Lee, J.S. (2006). Preschool teachers' shared beliefs about appropriate pedagogy for 4-year-olds. *Early Childhood Education Journal, 33*(6), 433–441.

Lee, J. (2010). Exploring kindergarten teachers' pedagogical content knowledge of mathematics. *International Journal of Early Childhood, 42*(1), 27–41.

Lee, J.S., & Ginsburg, H.P. (2007). Preschool teachers' beliefs about appropriate early literacy and mathematics education for low- and middle-socioeconomic status children. *Early Education and Development, 18*(1), 111–143.

Lee, Y.-S., Hartman, G., Pappas, S., Chiong, C., & Ginsburg, H.P. (2011). *Elaboration of the Early Mathematics Assessment System: The Spanish version and screen*. Paper presented at the Society for Research in Child Development, Montreal, Quebec, Canada.

Levine, S.C., Suriyakham, L.W., Rowe, M.L., Huttenlocher, J., & Gunderson, E. (2010). What counts in the development of young children's number knowledge? *Developmental Psychology, 46*(5), 1309–1319.

Lin, H.L., Lawrence, F.R., & Gorrell, J. (2003). Kindergarten teachers' views of children's readiness for school. *Early Childhood Research Quarterly, 18*(2), 225–237.

Maxwell, K., Lim, C.-I., & Early, D.M. (2006). *Early childhood teacher preparation programs in the United States: National report*. Chapel Hill: University of North Carolina, Frank Porter Graham (FPG) Child Development Institute.

McCray, J.S., & Chen, J. (2012). Pedagogical content knowledge for preschool mathematics: Construct validity of a new teacher interview. *Journal of Research in Childhood Education, 26*(3), 291–307.

National Association for the Education of Young Children & Fred Rogers Center. (2012). *Technology and interactive media as tools in early childhood programs serving children from birth through age 8.* Position Statement. Washington, DC: Author.

National Association for the Education of Young Children & National Council of Teachers of Mathematics. (2002). *Early childhood mathematics: Promoting good beginnings.* Washington, DC: Author. Retrieved from http://www.naeyc.org/files/naeyc/file/positions/psmath.pdf

National Association of Child Care Resource and Referral Agencies. (2012). *Leaving children to chance: 2012 update.* Retrieved from http://www.naccrra.org/about-child-care/state-child-care-licensing/training-requirements

National Council for Accreditation of Teacher Education. (2010). *Transforming teacher education through clinical practice: A national strategy to prepare effective teachers.* Washington, DC: Author.

National Council of Teachers of Mathematics. (2000). *Principles and standards for school mathematics.* Reston, VA: Author.

National Council of Teachers of Mathematics. (2006). *Curriculum focal points for prekindergarten through grade 8 mathematics: A quest for coherence.* Reston, VA: Author.

National Governors Association Center for Best Practices & Council of Chief State School Officers. (2010). *Common Core State Standards.* Washington, DC: Authors. Retrieved from http://www.corestandards.org

National Research Council. (2009). *Mathematics learning in early childhood: Paths toward excellence and equity.* Committee on Early Childhood Mathematics, C.T. Cross, T.A. Woods, & H. Schweingruber (Eds.). Center for Education, Division of Behavioral and Social Sciences and Education. Washington, DC: National Academies Press.

National Research Council. (2010). *Preparing teachers: Building evidence for sound policy.* Washington DC: National Academies Press.

Pianta, R.C., Howes, C., Burchinal, M., Bryant, D., Clifford, R.M., Early, D.M., & Barbarin, O. (2005). Features of pre-kindergarten programs, classrooms, and teachers: Prediction of observed classroom quality and teacher-child interactions. *Applied Developmental Science, 9*(3), 144–159.

Piotrkowski, C.S., Botsko, M., & Matthews, E. (2001). Parents' and teachers' beliefs about children's school readiness in a high-need community. *Early Childhood Research Quarterly, 15*(4), 537–558.

QRIS National Learning Network (2014). Current status of QRIS in states. Retrieved from: http://www.qrisnetwork.org/sites/all/files/maps/QRIS%20Map,%20QRIS%20National%20Learning%20Network,%20www.qrisnetwork.org%20%5BRevised%20February%202014%5D.pdf

Reardon, S.F. (2011). The widening academic achievement gap between the rich and poor: New evidence and possible explanations. In G.J. Duncan & R.J. Murnane (Eds.), *Whither opportunity?: Rising inequality, schools, and children's life chances* (pp. 91–115). New York, NY: Russell Sage Foundation.

Sarama, J., & DiBiase, A.-M. (with Clements, D.H., & Spitler, M.E.). (2004). The professional development challenge in preschool mathematics. In D.H. Clements, J. Sarama, & A.-M. DiBiase (Eds.), *Engaging young children in mathematics: Standards for early childhood mathematics education* (pp. 415–446). Mahwah, NJ: Lawrence Erlbaum Associates.

Schram, P., Wilcox, S.K., Lapan, G., & Lanier, P. (1988). Changing preservice teachers beliefs about mathematics education. In C.A. Mahers, G.A. Goldin, & R.B. Davis (Eds.), *Proceedings of PME-NA 11* (Vol. 1, pp. 296–302). New Brunswick, NJ: Rutgers University Press.

Scott-Little, C., Kagan, S.L., Reid, J.L., & Castillo, E. (2011, December). *Early mathematics standards in the United States: Understanding their content.* Prepared for the Heising-Simons Foundation. New York, NY: National Center for Children and Families.

Smith, K.H. (1998). *The construction of a survey of pedagogical content knowledge in early childhood mathematics.* Unpublished manuscript.

Smith, K.H. (2000). *Early childhood teachers' pedagogical knowledge in mathematics: A quantitative study.* (Unpublished doctoral dissertation). Georgia State University, Atlanta.

Stipek, D.J., & Byler, P. (1997). Early childhood teachers: Do they practice what they preach? *Early Childhood Research Quarterly, 12*(3), 305–325.

Stipek, D.J., Givvin, K.B., Salmon, J.M., & MacGyvers, V.L. (2001). Teachers' beliefs and practices related to mathematics instruction. *Teaching and Teacher Education, 17,* 213–226.

Shulman, L.S. (1986). Those who understand: Knowledge growth in teaching. *Educational Researcher, 15*(2), 4–14.

Thompson, A.G. (1992). Teachers' beliefs and conceptions: A synthesis of the research. In D.A. Grouws (Ed.), *Handbook of research on mathematics teaching and learning* (pp. 127–146). New York, NY: Macmillan.

Tout, K. (2013, March). *Quality improvement supports in QRIS: Evidence and practice.* Paper presented at the Quality Improvement Meeting, National Center for Research on Early Childhood Education (NCRECE), Washington, DC.

Tout, K., Starr, R., Soli, M., Moodie, S., Kirby, G., & Boller, K. (2010). *Compendium of quality rating systems and evaluations.* Washington, DC: U.S. Department of Health and Human Services, Administration for Children and Families, Office of Planning, Research and Evaluation.

Trice, A.D., & Ogden, E.D. (1987). Correlates of mathematics anxiety in first year elementary school teachers. *Educational Research Quarterly, 11*(3), 2–4.

U.S. Department of Education, Policy and Program Studies Service. (2010). *Toward the identification of features of effective professional development for early childhood educators: Literature review.* Washington, DC: Author.

Varol, F., Farran, D.C., Bilbrey, C., Vorhaus, E.A., & Hofer, K.G. (2012). Improving mathematics instruction for early childhood teachers: Professional development components that work. *NHSA Dialog: A Research-to-Practice Journal for the Early Childhood Field, 15*(1), 24–40.

Wartella, E., Schomburg, R.L., Lauricella, A.R., Robb, M., & Flynn, R. (2010). *Technology in the lives of teachers and classrooms: Survey of classroom teachers and family child care providers.* Latrobe, PA: Fred Rogers Center, St. Vincent College.

Wasik, B. (2008). When fewer is more: Small groups in early childhood classrooms. *Early Childhood Education Journal, 35*(6), 516–521.

Woods, T.A. (2011). *What can we do to support all children's math learning? Supporting teachers' math knowledge and practice.* Unpublished manuscript.

3

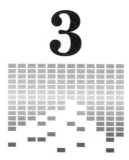

Young Children's Mathematical Minds

(Almost) All About Ben

Herbert P. Ginsburg[1]

You cannot do effective professional development in early math education or teach the subject well unless you understand children's mathematical minds. They are remarkable, surprising, fascinating, and amusing. Hence this chapter, which is (almost) all about Ben.

I try to bring to life the story of children's mathematical development by presenting and interpreting excerpts of extensive video interviews with Ben at ages 3, 4, and 5. For those designing or providing professional development, his adventures with mathematics offer valuable lessons about what young children know and need to know about math. The videos also portray how I interact with Ben to learn what he knows. My technique of clinical interviewing—flexible questioning of the child—offers lessons about assessment and teaching. As an added bonus, we also see that the mathematical ideas with which children engage are (almost) as interesting as the children themselves. At the end of the chapter, I discuss implications for professional development. But first some background.

EVERYDAY MATH

We tend to think of math as *only* an academic subject—painful for many, dry for some, and fascinating for others. But math is more than what is taught, often poorly, in school. A living, everyday form of math exists in the lives and minds of children from the ages of 2 to 4 before the onset of formal schooling. Here is a simple list, the accuracy of which you can verify from your experience. Young children

- Argue about portions of food: "Why does she have more than I do? It's not fair!" (meaning equal amounts, an idea forming the foundation of justice)

- Play board games, such as Sorry!: "I can go 12 spaces. I hope you have to go back 4 spaces."
- Read wonderful storybooks, such as *Goldilocks and the Three Bears,* introducing the concept of small, medium, and big, each of which gets a corresponding amount of porridge
- Read number books, such as *Anno's Counting Book*
- Look for the numbers on the bus: "We have to take the number 3 bus"
- Build block structures symmetrical in three dimensions (such as a castle with turrets on both sides)
- Count along with the Count on *Sesame Street*
- Enjoy counting: "I can count real high!"
- Use touch-screen devices and computers from the age of 2 to play with math apps
- Watch the floor numbers on elevators and sometimes press all the buttons
- Learn to write numerals
- Learn the patterns in music and dance
- Sing songs such as "The Wheels on the Bus"
- Ask "What is the biggest number?"
- Play "10, 9, 8 . . . blast off"
- Deliver the ultimate insult: "I'm big; you're just a baby"

For the most part, children's everyday math is not written or formal. They do learn to read and write a few numbers and may even get math instruction from computer apps and from parents, some of whom unfortunately use workbooks to drill their children in written math. But children's everyday math entails mostly informal components, some artistic (patterns in music), some motivated by greed or rivalry ("I'm bigger than you"), some playful and competitive (Sorry!), and some originating in intellectual curiosity (the largest number).[2] Many parents and educators (a general term used in this chapter to refer to teachers, teaching assistants, caregivers, and others involved in early education) are unaware of children's everyday math—or at least its major features. This should not be surprising. After all, apart from a small and inbred group of cognitive researchers, most psychologists are also unaware of everyday math.

So it is clear that young children everywhere practice everyday math. But we need to understand it in some depth. What do children really know about patterns, numbers, and relations (e.g., the correlation between bear size and amount of porridge)? Is children's knowledge abstract and substantive, or do they only memorize a few math words? And for education, there is one fundamental question: How can we use what is known about children's everyday math to inform professional development, both in-service and preservice, and ultimately the practice of early childhood math education?

MEET BEN

I interviewed Ben on or very close to his third, fourth, and fifth birthdays. His parents brought him into a room on the Columbia University campus, where Ben and the interviewer (yours truly or Janet Eisenband, my doctoral student at the time) sat at a table to do various "games" as a videographer[3] captured the interactions. Ben's parents stayed in the room, observing and occasionally interacting with him. As you will see, the atmosphere did not involve high-stakes, pressured testing; Ben clearly enjoyed the interactions. In fact, his father told me that after the first session, Ben looked forward to playing again with Dr. Ginsboo.

Here is how our session began when he was 3 years and a day old. At the very outset, Ben puts some chips on the table, commenting with some excitement, "Look, they are all yellow." He then turns them all over one at a time, so that they are now all red, and is delighted at the outcome. You might even say that he was joyful.

It is marvelous to observe excitement such as this, and our immediate response (or at least mine) is that Ben is enormously cute. He is, but we cannot leave it at that. Episodes such as Video 3.1: Ben with Yellow and Red Chips, Age 3 (see the About the Ancillary Materials page for details) can teach us that this form of assessment—the clinical interview—can be an exciting and enjoyable experience for child and adult alike, although Ben did get temporarily bored after about 34 minutes. The interviews can provide important insights into children's thinking. For example, I would argue that Ben was not excited about playing with the chips themselves. Instead, as a free agent, he decided to play with ideas about the chips and what he could do with them. Ben seemed fascinated with the idea of *all,* perhaps in contrast to the idea of *some*. And he was even more excited by the fact he could *transform* all yellow into all red. Why? I am not sure. But it is true that the ideas of *all* and *some* and *transformation* are fundamental in mathematics, logic, and everyday life, and it was good that Ben was exploring them and found them so interesting. And all this in the first 19 seconds!

But then the following question immediately arises: How typical is Ben? The answer is that he is atypical, typical, and typically atypical. He is *atypical* in that he is a privileged child whose intellectual functioning is probably in advance of his peers, although I never gave him a test to investigate that issue, and I am unaware of results from relevant tests he may have taken in school. He is *typical* in that everything he does illustrates the kind of mathematical development that virtually all children experience. For example, Ben may learn to count at a very young age, but in doing so, he displays the same kinds of struggles, errors, and triumphs as do other children. He is *typically atypical* in that all children are atypical—unique—in one way or another. The adult's job is to use general knowledge of mathematical development to understand the individual child's mind. Ben's example can help us do that.

As you watch the videos, try to do several things. Most important, observe carefully and try to figure out the meaning of Ben's behavior. It is easy to see, for example, that he is counting little bears. But ask yourself: Why? What are the concepts with which the child is grappling? What is the child thinking?

Try to get beneath the behavioral surface to learn about the ideas and thinking underneath.

Second, focus on the adult's role. Watch how the clinical interviewer operates, first asking open-ended questions, but then posing new, probing questions designed to examine hypotheses about underlying mathematical ideas and strategies. Observe how the interviewer sometimes challenges Ben's ideas to find out what he really thinks. Also, see how the interviewer sometimes uses a form of indirect hints or scaffolding to determine whether the child is capable of a solution with a little indirect adult help. The interview is very, very different from standardized testing and didactic teaching and can make a significant contribution to early childhood education (and parenting, too). By contrast with traditional methods of testing and teaching, the interview is an interaction in which the child encounters an adult who joins him in undertaking a fascinating and amusing intellectual adventure.[4] In Chapter 4, Juanita Copley argues that developing proficiency in interviewing and other forms of assessment should be a major goal for professional development. In Chapter 6, Michael D. Preston discusses the teaching of clinical interviewing in preservice settings.

COUNTING WORDS AND COUNTING THINGS

Spoken and written number words permeate children's everyday world. "Here are two cookies for you." "Sesame Street is on channel 13." "Press the 7 button. We have to go to the top floor." "You can't have all three blocks. Give one to your brother." "Wow, you weigh 36 pounds!" "We're going to read the story about the three pigs." "One, two, buckle your shoe. Three, four, shut the door." "How high can you count?" "Two, four, six, eight. Who do we appreciate?"

Early on, children take a great interest in number words and ideas. Some children love counting as high as they can, like grown-ups. They may even be interested in the name of the "biggest" number.

At the age of 3, Ben had this unsatisfactory conversation with his mother:

    Ben: Mama, but what is the *last* number? The one at the end.

    Christine: There is no last number. You can count forever without reaching the end.

    Ben: Mama, are you sure you're listening to me?!

Children's spontaneous interest in number words is useful because fluency in counting is important for children's computation and math achievement in general. But counting is not as simple or dull as it may appear. Consider two aspects of counting, saying the *counting words* (when they do not refer to things) and *counting things* (enumerating).

Counting Words

The first 12 English counting words are completely arbitrary. The first in the series could be "funf," the fifth could be "uno," and the twelfth could be "pineapple." As a result, children have to engage in rote learning to memorize the sequence, just as they memorize the alphabet or learn the words to a song ("My country tizofthee").

Adults may not remember how hard it is to learn the sequence of counting words, which from children's point of view are an aural blur. Watch and listen carefully to the four described videos and then try to repeat what the speaker[5] says. I will bet you cannot even isolate the individual words that need to be memorized.

In Video 3.2a: Counting in Farsi, 1–40 (go online to view the video), Azi counts from 1 to 40 in Farsi, spoken at a relatively fast pace. Now Video 3.2b: Counting in Farsi Slowly, 1–19 (go online to view the video) is a slower version of counting from 1 to 19, which you will undoubtedly have no trouble replicating.

Now that you did so well on that, Video 3.2c. Counting in Farsi Slowly, 1–40 (go online to view the video) shows Azi again counting from 1 to 40 slowly.

You are an excellent counter! Now you can easily count to 40 very quickly, just like Azi does in Video 3.2d: Counting in Farsi Faster, 1–40 (go online to view the video).

Well, the truth is that your Farsi counting efforts were probably pretty pathetic. But children do eventually memorize the counting words and at the same time learn some interesting concepts. Video 3.3: Ben Counting, Age 3 (go online to view the video) shows Ben at age 3. Janet asks him to check the number of toy bears. This of course is an enumeration problem, but Ben turns it into an opportunity to display his knowledge of counting numbers beyond the number of plastic bears.

A year later, in Video 3.4: Ben Counting, Age 4 (go online to view the video), Ben, now 4, counts (with his hand in his mouth) up to 17, skipping only the number 16. After that, with a little help, he makes it up to 20, which he seems to think is the last number.

Next, in Video 3.5: Ben Playing "Catch My Mistake" Game, Age 3 (go online to view the video), Ben plays the "catch my mistake" game for the first time after he has just turned 3. He can tell when a mistake occurs, and he tries with some success to vocalize the problem.

Video 3.6: Ben Playing "Catch My Mistake" Game, Age 4 (go online to view the video) shows Ben again at age 4. This time, when Ben plays the "catch the mistake" game, he is clearly able to verbally identify almost all mistakes.

At both ages, he clearly enjoys turning the tables on an adult and correcting the habitual corrector. The activity is a lot of fun but also shows that Ben believes several general principles about numbers:

- It is not permitted to skip a number in the sequence. He knows this at age 3, and at 4, he can provide what is missing.

- It is not permitted to say a number twice. At age 3, he is particularly worried about how the repeated "three" will multiply, so to speak, in an unmanageable way.

- Numbers are different from colors.

Other children express another important rule: "You have to start with one!"

This means that Ben, along with other children, is not only memorizing the numbers but also thinking about what he is memorizing—even at the beginning of his 3rd year. He has figured out, probably by himself, that the category of numbers is different from the category of colors (would anyone have ever explicitly taught him that idea?) and that there are rules for counting: Do not repeat a number, and do not skip any. Almost all children learn the start-from-one rule, which can be cumbersome, for example, as in the case of adding 1 to 45.

At 3 years old, Ben is even practicing to be a cognitive psychologist when he decides to play the mistake game with Janet as the experimental subject (see Video 3.7: Ben Deciding to Be the Experimental Subject [go online to view the video]). He makes the deliberate mistake "1, 2, 20" and is pleased with himself and the response he receives.

After memorizing the first 12 numbers, learning that numbers are different from other categories such as colors and acquiring rules about skipping and repeating and starting from one, English-speaking children are faced with a challenge: The words from 13 to 19 are difficult to learn. One approach is to memorize them, which would bring to 19 the total number of number words to be learned in this way. The other approach is to learn the underlying pattern—namely, that the unit word (at least, "three" through "nine") is followed by "teen," which derives from "ten." Thirteen is a form of "three-ten;" fourteen is a form of "four-ten;" and so forth until "nine-ten." But the pattern may be hard to detect, and the words are strange too. "Thir" stands for "three," and "teen" for "ten." Fourteen is easy, but "fif" in fifteen stands for "five."

At age 5, Ben has no difficulty with the "teen" words from 13 to 19; he rattles off the numbers from 1 to 29 (see Video 3.8: Ben Counting from 1 to 29, Age 5 [go online to view the video]).

But when children get to 20, they have to learn a new rule for generating numbers. The rule is simple and underlies our base-10 system of numbers: You first begin with the appropriate "tens" word (20, 30, 40, and so forth through 90), and then you simply append to it the numbers from one to nine. Similar to the teens words (13 and friends), the tens words are a little odd: Twenty is not much like two-ten (or "two-ty"), and the first syllables of "thirty" and "fifty" sound different from three and five, respectively. But more important, this is the first time children grapple with the idea that we can think of numbers as groups of tens and other units. The idea also underlies our "place value" system for writing numbers, in which the three in 36 refers to 3 tens and the six to 6 units.

In Video 3.9: Ben Counting after 30, Age 5 (go online to view the video), Ben has trouble with the new rule. Instead of 30, he says 50 and goes on from there, correctly, to 59 when he stretches out the "nine" as he searches for 60. As Ben is trying to figure out what number comes after 59, I pause to let him think. In Ben's case, the silence was not only golden but also effective.

A digression about interviewing technique may be useful. The decision not to say anything is challenging and risky for an interviewer. The silence may encourage the child to think or it may make him unhappy if he cannot solve the problem. Sometimes teachers (I use this term to refer to those who work directly with children in classrooms, child care centers, and the like) and

> People from China and other East Asian countries have an easier time learning these numbers because their languages are mathematically well structured (Miller & Parades, 1996). In Chinese, they say the equivalent of ten, ten-one, ten-two, and so forth (instead of the obscure eleven and twelve), and then they switch to two-ten, two-ten-one, and so forth in a typical sequence to nine-ten-nine. The words are clear, and the pattern is coherent, reflecting the base-10 structure of the numbers.
>
> This elegant structure gives Chinese-speaking children an advantage when they later learn to "carry" in school mathematics. Suppose you have to add two-ten eight (28) and three-ten-seven (37). Adding the units, you get one-ten-five and then "carry" that one-ten so that you can add it to the two-ten and the three-ten. It is conceptually clear: The meaning of carrying is conveyed by the number words themselves.

parents, too, do not wait long enough before providing the correct answer, perhaps because they do not want the child to be wrong at all (and also perhaps because we adults like to reinforce our status as the knowers and tellers). Of course, we do not want to encourage incorrect answers, but we want to get across the idea that it is important to think and to check and that getting the right answer is not the only goal. Providing correct answers at the slightest sign of error is a kind of occupational hazard for both teachers and parents.

But back to Ben.

After a while, in Video 3.10: Ben Counting to 79, Age 5, he finds 60 and goes on to 79, which he says is the last number he can count to. But I am not so sure and give him some simple hints.

In Video 3.11: Ben Counting Past 79 with Help, Age 5 (go online to view the video), I start by asking Ben what comes after seven. My intention is to draw a parallel between the unit and tens numbers. Ben gets it, and then rattles off the numbers to 89. But there he gets stuck, and I need to intervene. At first, I try the indirect method of asking him to think about what comes after eight, but he cannot make the connection between 9 (which of course he knew) and 90. Given this, I then more or less give him the answer and then throw in 100 as a bonus.

Was I interviewing or teaching him? The answer is yes to both. At first the interview begins in a very open-ended way, for example, with a question such as "How high can you count?" But then, as we have seen, the interviewer may introduce hints that are intended to scaffold the child—that is, to lead him along, very gently, to discover the answer for himself, seeing, for example, that he can use the unit numbers to infer the tens. And if that does not work, then some explicit telling of the answer can be useful.

In Video 3.12: Interviewing to Discuss What Children Know/Teach (go online to view the video), I extended the teaching/learning method even further.

I begin by asking Ben what number comes after 123. He correctly replies 124. He struggles to find the number after 129, so I prompt him by asking what comes after 29 and mouth the "th" sound as well. Ben correctly says 30.

I then ask what comes after 139. When Ben struggles, I ask him what comes after 3. When he says 4, I ask again what comes after 139. He says 40, and I correct him by saying 140. We then count by tens together, though Ben struggles somewhat with the sequence. I refer him back to the sequence of numbers 1–10 to move him along. My strategy is to help Ben use his existing knowledge of the lower numbers to construct the higher ones.

This episode illustrates how interviewing to discover what children know evolves into teaching that draws on this insight so as to teach them something relatively "new"—that is, what they almost know. This is what "formative" assessment intends to do: The teacher tries to understand the child's thinking and then uses that knowledge to shape instruction.

To summarize, learning to count is not simple, at least in English. Children must memorize the first 12 numbers and along the way learn the rules about repetition, skipping, and the abstract concept that numbers are a special type of word (as opposed to, for example, the category of colors). Then children learn what turns out to be a backward pattern for the numbers from 13 to 19, with some strange-sounding morphemes (meaningful parts of words) such as "fif" and "thir." And next they have to unlearn the backward pattern in order to use the sensible base-10 pattern from 20 to 99 (although it too has some odd words such as "twenty" and "thirty"). If children have trouble learning to count, send complaints to the inventors of the English language.

Why is it important for providers of professional development and educators in general to know all this? One reason is to understand that even the apparently simple task of learning the counting words may engage young children in abstract mathematical thinking. Educators need to appreciate that young children are not simply concrete thinkers. A second reason is that, despite obstacles imposed by the English language, counting can be a very important pattern detection activity, which is fundamental to mathematical thinking. Third, the pattern to be detected is essential: the base-10 system that allows an orderly and consistent construction of numbers of increasing magnitudes. And finally, close observation of the clinical interview method illustrates both the process of formative assessment and how it can be used to guide teaching.

Counting Things

Knowing the counting words, children can apply them to the task of counting things to find out "how many." We call this activity "enumeration" so as not to confuse it with the other form of counting—namely, counting as saying the number words. Enumeration requires knowing the number words, but this knowledge alone is not enough. Without important *ideas* about number, a child may count out loud to 20 but still may not be able to accurately enumerate a group of 5 things. If learning the counting words is complex, enumeration is even more abstract and cognitively demanding.

Suppose we have the collection of objects shown in Figure 3.1: We use words to designate each. Going from left to right, we call them "lion," "school bus," and perhaps "icky." If you refer to the school bus as "lion," people will think that you need some serious help.

Figure 3.1. Some things.

Suppose now that we count the objects. We point to the lion and say "one," to the school bus and say "two," and to the icky and say "three." And then we decide to count them again. This time we point to the icky first and say "one," to the school bus and say "two," and to the lion and say "three." If we do not count only left to right, the lion can be "three" and even "two." The number words are clearly not like names. Names designate specific categories, but number words are used differently. Neither "one" nor "three" is the name of the lion.

This can be very confusing for young children who are used to words referring to things (and are constantly trying to learn new ones). One of my graduate students asked a child to count. The child said, "One, two, three, five." When asked why "four" was missing, the child replied that *she* was four. She seemed to use the counting word as a name for herself.

So one of the first difficulties in enumeration is that counting words are used differently from ordinary names. We can refer to the animal as "lion" but not "tiger" or "football." But we can refer to it as "one," "two," and "three" (or indeed *any* counting number). We use the counting words differently from ordinary names, thus violating ordinary usage and confusing young children (and sometimes adults too, as I show later).

Why this odd usage? What do the counting words refer to? To understand the issue, consider the basic mathematical ideas that underlie counting.

We must say the number words in the proper order. "One" cannot come after "four."

By contrast, the order of counting things does not matter. You can start with icky or you can start with lion.

We have to count each object once and only once. If we use the word "one" to refer to the lion, we cannot use it to refer to the icky. I once interviewed a child who, when asked to count some chips, began by arranging them in a circle and then counted the objects over and over again as he looped around the circle (which is the worst way to arrange objects if you want to count them). I had to stop him in order to continue the interview.

We can count any discrete unit—dogs, imaginary unicorns, or even more imaginary ideas ("I had two ideas today").

Not only that, but we can count any combination of things. A group can include dogs, unicorns, and ideas, or anything else.

Physical arrangement is irrelevant for counting. You can count objects when they are scattered around or when they are in a line.

Likewise, the physical nature of objects does not matter. Every object, no matter what it is, is a unit for the purpose of counting. This idea must be very strange to little children. Suppose you find that one group has three ants and the other has three elephants. Ants are small and elephants are large. How can both groups be "three"?

And finally, the really big and difficult idea: Each number in the sequence does not refer to an object but to the number of objects (the "cardinal value") up to that point. Adult counting is so automatic that we sometimes do not appreciate the idea behind it.

Imagine that you see these pictures of a strange looking arctic creature in an exotic land called "Riverside Park." You want to find out the number of creatures in total: the basic issue of how many. You begin by drawing a line enclosing the first picture and say "one" because there is one creature, as in Figure 3.2.

Next, you draw a line around both and say "two" for the same reason: There are two creatures within. But when you point to the second object and say "two," that object itself is not "two." The word refers to the numerical value—the cardinal value of the set, the collection of two arctic monsters, as in Figure 3.3.

Then you draw a line around all creatures and say that there are "three." Again, three refers to the whole group, not to the third creature, as in Figure 3.4.

One way of thinking about this is that you start with a set of one, and then that set of one is included within the larger set of two, and then that set of two (which contains one) is included in the larger set of three, as deftly illustrated in Figure 3.5.

I will bet that many readers have not thought about counting things in this way—as sets within sets, always increasing by one. The idea is not simple, and it is therefore no surprise that children have trouble with it.

At the outset of the first interview in Video 3.13: Ben Displaying Variable Command of Enumeration, Age 3 (go online to view the video), Ben displays variable command of enumeration.

At first, given six "elephants," he counts to five. Not bad: only off by one. When the elephants are turned into "bananas," he again counts to five, very quickly, not bothering to count each with care. He spews out the number words very quickly, sometimes—as in the case of the yellow chips—not even looking at what he is pointing to. Yet he was close, only off by one, on the first two problems: That is pretty good for a child turning 3. He is also inconsistent. He gets

Figure 3.2. One creature.

Young Children's Mathematical Minds 63

the number three wrong, calling it two, but then when I take away one, he says two again and does not appear to be troubled by the inconsistency. And then at the end of the sequence, he succeeds in counting three elephants and four elephants, as he very carefully points to one at a time.

Later in the session, after about 25 minutes into the interview, an impressive amount of time for a 3-year-old, Janet gives Ben a slightly different enumeration task, asking him to produce a certain number of bears. He shows

Figure 3.3. Two creatures.

Figure 3.4. Three creatures.

Figure 3.5. Sets of creatures.

varying levels of success in Video 3.14: Ben Counting Bears, Age 3 (go online to view the video).

He has no problem with producing one and two bears (although he is concerned to select specific bears, perhaps because he has not mastered the rule that a number can refer to any object whatsoever).[6] But when asked to produce four bears, he puts out two and then three and finally four (although he referred to the last collection as "three"). Asked to check the last result, his counting goes berserk, as he spews out the number words from one to ten.

Now let us turn to Ben at age 4. At first, in Video 3.15: Ben Counting Chips, Age 4 (go online to view the video), he finds it easy to produce small numbers, three and five. Each time, he carefully slides one "banana" toward me and arranges the result in a straight line, carefully setting it apart from the remaining chips. Certainly during the past 12 months, he can produce numbers in a more deliberate and controlled manner.

Then we turn to a bigger problem—namely, 8 chips—in Video 3.16: Ben Counting 8 Chips, Age 4 (go online to view the video).

We see that he gets a little confused after putting out six chips. After all, producing larger numbers requires more working memory—mental space in which to carry out and monitor cognitive operations on pieces of information—than does production of a smaller number. As children get older, their working memory expands, as does their repertoire of strategies for dealing with the task at hand.

When I suggest that Ben check his result, he notices that the chips appear to resemble the numeral 4. That, of course, does not help him solve the problem, but it shows that he has numerals on his mind. After that, he tries again to put out eight chips, but this time is a little sloppy about where he puts the chips, failing to push them aside or place them in a straight line so that he would count each chip once and only once. The result is that he gets an incorrect answer. Furthermore, it is not even clear that he is finished at the end. His inflection indicates uncertainty, and I may have stopped him too soon.

In brief, Ben does well with a relatively large number, but he needs to perfect strategies for careful enumeration. One lesson for educators is that children may get wrong answers for many reasons, one of which is that they are sloppy and lack strategies that can guard against carelessness and overcome the limits of working memory.

The next important issue is whether Ben understands cardinality—the idea that the last number in the count sequence signifies the total amount. You might expect that at age 3 Ben does not understand it. To find out, I introduce an elephant cardinality problem in Video 3.17: Elephant Cardinality Problem, Age 3 (go online to view the video).

In this clip, I cover the four elephants with my hand and ask Ben how many elephants are under my hand. Ben does not seem to understand cardinality and each time has to recount the elephants to determine their number, even when it is only two. Also, it is interesting that as he counts the four elephants at the outset, Ben says something similar to "one, it's a two, it's a three, it's a four." This is very close to saying that each number is the name of that particular elephant with which it is paired.

What about Ben at 4 years? As can be seen in Video 3.18: Elephant Cardinality Problem, Age 4 (go online to view the video), he was certainly able to enumerate larger numbers than he could at age 3. To learn about his understanding of cardinality, I cover up the chips that he had just counted and ask how many are under the sheet of paper.

After Ben says "zero," apparently referring to the number of chips on top of the paper, I remind him again of what he had previously said—namely, that there were eight chips.[7] But he was overwhelmed and gave up. As before, we see that Ben has yet to fully understand cardinality.

Next, in Video 3.19: Ben Counting 5 Apples, Age 4 (go online to view the video), I give Ben Piaget's classic "conservation of number" problem: Will he see that the number remains the same even if you simply rearrange the objects into a new and different-looking array? This task would seem to be easier than the hiding problem because Ben can now see everything that happens to the objects—nothing is hidden. I asked him to give me five apples and he did.

But to determine the number, Ben had to count them all over again, just as he did when they were hidden. The total is unclear to him when the coins are rearranged. Next, in Video 3.20: Ben Counting 3 Apples, Age 4 (go online to view the video), I decide to give him a very small number, three.

Ben is able to recognize that three apples remain three in number even when they are spread out or returned to their original position. He justifies his answer by counting. When I place a piece of paper over the same three apples and ask how many there are, Ben is again overwhelmed and gives up.

In brief, learning enumeration is complex and difficult. It requires a new use of words to refer to concepts of number (such as "fiveness"), not to things. It involves basic and deep mathematical ideas, such as the irrelevance of physical attributes for number: Three little things are "more" than two huge things, even if the latter are larger than the former. Children take a long time to master the ideas and the strategies required for enumeration, such as carefully counting each thing once and only once so as to reduce the demands on memory.

Why is this valuable for educators to know? One reason is to appreciate that early math is not baby math: Young children have to grapple with deep mathematical ideas, which educators need to understand. A second reason is to realize that young children employ ideas and strategies to get their answers, and so it is important for educators to know what these strategies are and how to promote more efficient ones. A third reason is to understand that the strategy is as important as, if not more important than, the answer. A child can use very good strategies that result in wrong answers because of a little bit of sloppiness. A child can also get a correct answer but not know what it means, as in the case of the child who counts three objects correctly but does not know that the last number counted indicates the total amount, the cardinal number.

EVERYDAY ADDITION AND SUBTRACTION

One kind of addition grows directly out of enumeration. You can think of enumeration—determining the number of things—as adding by ones. You start first with one and add one to it to get two and then add one more to that to get three and so forth. When children figure out *how many objects there are in a set,* they at first count one by one. But then they sometimes take short cuts. Suppose a child sees a bunch of things in front of her. She notices that there are three over here and then some more. Her solution is to count, "Three . . . four, five." So in determining *how many,* children independently discover addition.

Children can work with several different kinds of addition. In Video 3.21: Ben Adding Chips (go online to view the video), I ask Ben to determine the number of objects in two different groups of chips. He got the wrong number just a moment before, so I remind him to count very carefully.

He gets the right answer, five, and then, without counting, "sees"[8] that there are three chips in the other pile. When I ask Ben how many there are altogether,[9] he announces that he has to count them, which he does, starting with one, getting the right answer. At the outset then, adding is simply combining two groups (a union of sets) and counting all the units to get the sum. Of course, there is an easier way to do it—namely, to count on from the larger number. Ben already knew that one group had five because he had counted it. Why not simply count on three more from the original five to get the answer? Older children do this, often discovering the method on their own.

Next, in Video 3.22: Ben Adding Pirate Coins (go online to view the video), Ben gets extremely excited when he hears that we are going to play a new game, invented by Zur and Gelman (2004). He does not tolerate too much chit-chat before he asks, "So what do we have to do?" I hide nine pirate coins, one at a time, in the pirate's secret bag. Then I tell him that I added two more. His job is to figure out how many there are altogether.

At first, after whispering the numbers from 1 to 9, Ben again says, "I give up." But then, just a little scaffolding helps him get the answer. He first asks what number comes after nine, and when given ten, he counts on by one more to get the correct answer. Though at first he proclaimed ignorance, he knows enough to ask for the number after nine. And right after that, he counted on to get the next and final number.[10] This means that he has a basic concept of addition as counting on from the first addend—at least when he gets a little help from an adult. But did we not see that in the addition-as-putting-together task, he counted all to get the answer? That is true, but here the task itself, in which both groups of chips are hidden in the bag, channels his thinking into the idea of counting on. In other words, because he could not see the chips and would find it extremely difficult to count their images in his mind, he had almost no choice except to count on. So tasks matter: They support or discourage different strategies. This is as important for teachers as it should be for testers.

Note this about assessment: Contrast this informal, supportive testing situation with a standard test in which a child is given only one chance to get the answer. On the initial "test," Ben's initial response was to capitulate, which would have earned him a score of 0, presumably indicating ignorance of addition. This misguided interpretation does not show that testing itself is bad. It does show that bad testing is bad, and bad testing is often standardized testing, especially for young children. Flexible testing is likely to be more accurate and quite enjoyable for young children.

Video 3.23: Ben Counting Chips (go online to view the video) is a slightly different example of Ben's understanding of addition. This time, I place three chips under a piece of paper and then two more. He cannot see the result. I ask him to put the same number of chips under his paper. This time, he does not have to tell me the answer; he just has to produce the right number. And he does, collecting five coins and placing a piece of paper over them.

The numbers in this task are much smaller than those in the previous Zur and Gelman problem. What is remarkable here is not the fact that he easily gets the sum but that he is quite aware of how to solve the problem and can describe the needed method in advance. He says, "We have to go three and then two more." And not only that, but he proclaims with some joy that the two sets have the "same number."

Ben and I continued to play various "games" for about 50 minutes in total. We were getting ready to wrap up when Ben decided it would be fun (he literally jumped up in his seat) to play another game with Janet. In Video 3.24: Ben Playing a Game with Janet (go online to view the video), she gives Ben another pirate money task, a very difficult one.

She first gives both of them the same small number of pirate coins, four. Then she hides her four and asks Ben to watch as she takes one of his coins and places it with hers. So now he has three and she has five. I thought Janet was going to ask how many coins are in her hiding place. But she posed a much more difficult problem that involved equalizing.

She asked Ben to select from the larger pile enough coins to make his collection have the same number as hers. If Janet's and Ben's coins are initially the same number, and if Janet takes away one from Ben and adds that one to her collection, then how many coins does Ben need to restore the initial equivalence? Ben's response surprised me.

He was able to solve two problems of this type, at least with small numbers. He could have solved the problems by adding everything in his head as he went along. In other words, in the first problem, he sees that Janet takes 1 more for herself, adds 3 + 1, and gets 4. He also sees that he lost 1, subtracts 1 from 3, and gets 2. Then he subtracts 2 from 4 to find that Janet has 2 more than he does, and that is the amount needed to make them have the same number once again. Another more general way to solve it is this: She took away one from me and gave it to herself. That means that she has one more than she started with and that I have one fewer. So I need to give myself one to make up for the one she took and another one to make up for the extra one that she got.

Video 3.25: Ben Counting Bears, Age 5 (go online to view the video) is one more example, this time from Ben just after he has turned 5. We first establish that there are six toy bears under a piece of paper. Then as he watches, I put two more under the paper and ask him how many there are altogether. Recall that I showed him a very similar problem the previous year.

At first, he is quiet, apparently whispering some numbers to himself. Then he pops up and says a triumphant "eight!" Asked how he knew that, Ben says that four and four make eight and also that he counted on two more from six. I suspect he did that first and then remembered that four and four is eight. And then he volunteers that "if you make two more, it's ten." When I asked him a series of questions involving the addition of 2, he got up to 14 and then made a mistake. Despite this, it is clear that at 5 years old, he is more at ease with addition problems, seems to understand the equivalence between 6 + 2 and 4 + 4, and even creates and solves a new addition problem that begins with his previous answer.

In brief, these examples show that in a sense, young children already know a good deal about addition before they get to elementary school. They can figure out what happens when you add by combining two sets and what happens when you start with a set and add more to it. They can deal with some abstractions, such as doing the Zur and Gelman task, where one set is completely hidden. They can invent and solve problems on their own. They may even be able to describe their own strategies,[11] although many children are less introspective and expressive than Ben, who may be a bit precocious. Finally, the tasks you give a child make a huge difference. Asking the child to combine two visible sets privileges counting one by one, whereas the Zur and Gelman task channels thinking into the more efficient counting-on strategy.

Why is all this important for professional development providers (and educators in general) to know? One reason is to understand that children can learn a great deal through their own experience, without teaching. You need to give them opportunities to explore and to practice. A second reason is to appreciate that children may be more competent in mathematical thinking than we ordinarily imagine. This should come as no surprise. If we do not pose challenging, carefully designed problems, we will never learn that children can solve them.[12] A third reason is to realize that math education must encourage language, explanation, and thoughtfulness. In a sense, math education is literacy education.

WRITTEN, SYMBOLIC MATH

Eventually, children need to learn formal, written, symbolic math, which is one of humankind's greatest intellectual achievements, developed over centuries in many cultures (Hindu, Arab, and European). It is both intellectually challenging and practically useful. In contrast to children's everyday math, formal math is abstract, written, systematic, explicitly defined by the professional community, coherent, and organized. Mathematicians also find beauty in it. Formal mathematics is more powerful and useful for many purposes than is everyday math, and there is no doubt that children should learn the former.

As we have seen, young children are off to a good start: They enjoy math and develop an impressive array of concepts and skills. The educational goal is to draw on young children's positive motivation and everyday math to help them begin to understand the formal system. Another way to put it is that schooling should not teach children to fear math by the second or third grade, which happens all too often under present conditions.

Even during the preschool years, math education should begin to introduce written symbols such as the numerals and the +, −, and = signs. The highly influential Common Core State Standards (National Governors Association Center for Best Practices & Council of Chief State School Officers, 2010) recommend, for example, that students should be able to

> compare two numbers between 1 and 10 presented as written numerals . . . Decompose numbers less than or equal to 10 into pairs in more than one way, e.g., by using objects or drawings, and record each decomposition by a drawing or equation (e.g., 5 = 2 + 3 and 5 = 4 + 1). (p. 11)

The Common Core places diminishing emphasis on objects and drawings for Grades 1 and above, so that after kindergarten, written symbols make up a large portion of instruction.

But many children have considerable difficulty in learning formal math. Why? After all, their everyday math is reasonably competent, as we have seen. Yet the struggle with symbols begins as soon as they are introduced.

The teaching of the "equals" sign (=) and other basic written symbols illustrates the challenge that teachers face and that professional development must address. In kindergarten or first grade, children begin to learn simple arithmetic statements such as 3 + 2 = 5. The simplest task is learning how to write the numerals correctly. Children often write 2 as 5, for example, because they have difficulty in orientating the numerals, one of which faces left and the other right. After all, if a cat can face left or right and still be a cat, why should orientation matter for numerals or letters? But not to worry: Children eventually learn the correct orientation for numerals; the issue usually resolves itself.

The more serious challenge is to help children learn the *meaning* of the symbols. They need to know that 3, 2, and 5 are symbols for numbers and that + refers to addition. They also need to know the meaning of =. Teachers and textbooks generally say that = refers to "equivalence": The numbers on both sides of the = sign are the same. Thus 3 + 2 is equivalent in number to 5. In this view, 3 + 2 is one way of "talking about" or "naming" the quantity "five," and the numeral 5 is another. Equivalence is an important idea not only for basic numbers but also for algebra.

But what do children understand of these written symbols? Here is an example from Toby, a first grader, almost 6 years of age. I ask her to write "three plus four equals seven,"[13] and she writes down the symbols correctly. Then I ask her, "Can you write five equals two plus three?" She writes 5 = 2 + 3 but seems hesitant. So I ask whether it is "right" to do it that way, and she shakes her head to indicate the negative and adds that the numbers need to be reversed as 2 + 3 = 5.

Many other children join Toby in the mistaken belief that the only correct way to write addition is in the form a + b = c. I followed up to learn why.

Interviewer (I): Can you tell me, like in this one right here, we have three plus four equals seven, what does plus mean?

Toby (T): I'm not sure . . . I don't know.

I: I mean, does plus tell you do something . . . what is it all about? [A pause; she shrugs.] Not sure? What about equals? What does equals mean?

T: It tells you, um, I'm not sure about this . . . I think . . . it tells you three plus four, three plus four, so it's telling you that, um, I think, the, um, the end is coming up . . . The end.

I: The end is coming up . . . what do you mean the end is coming up?

T: Like, if you have equals, and so you have seven, then. So if you do three plus four equals seven, that would be right.

I: That would be right, so equals means something is coming up... like the *answer*. [We both laughed.]

In brief, Toby has an "action" interpretation of addition. For her, addition is an operation, and number sentences, which must be written from left to right, tell you what to do. When she sees a + b =, she knows that adding is required. She cannot say what + means, but clearly it sends her the message that she ought to add. Furthermore, she maintains that = means "The end is coming up." If + shouts "Add up those numbers," then = screams "Put the answer here." Toby's interpretation of = is not at all unusual, although most children say that = means "makes."

Why should educators care about these trivial mistakes? One reason to care is that students' view of = as an operation persists into middle school algebra, where the idea of equivalence is crucial. It is therefore important to teach the idea of equivalence and symbols effectively from the onset of schooling.[14]

Another reason is more fundamental and general. The analysis of Toby's thinking suggests that in school, children may not learn what teachers intend to teach. For most children, = essentially means "Operate to get an answer," whereas teachers believe that = means that both sides of the expression are equivalent. This indicates that many teachers are not teaching what they think they are teaching, and many students do not learn what teachers think they are learning. Teachers need to understand that children (and adults) assimilate what is taught into what they know. So in introducing symbolic math from the outset, teachers need to base instruction on an accurate assessment of what children bring to the task of learning mathematics. They need to relate the symbolic math to children's everyday math so that children expand their intuitive ideas and understand the adult's math, the formal, written system. When this is not done, children often end up seeing mathematical symbols as meaningless and arbitrary, as "academic," in the sense of meaningless, arbitrary, imposed from above, and not relevant to one's interests or informal abilities.

WHAT ALL THIS MEANS FOR PROFESSIONAL DEVELOPMENT

I have used carefully selected episodes of interviews with Ben to illustrate some—but not all—of the fascinating findings and insights that psychological researchers have produced over roughly the past 40 years. The portrait of mathematical development thus revealed has important implications for professional development, which should deal with two major topics. The first is what teachers need to understand, and the second is the question of how teachers can implement the new vision of early childhood math education.

What Teachers Need to Understand

First, children's mathematical development is rich and complex. We have seen that even 3-year-olds engage in mathematical thinking and have an interest in abstract issues such as the size of the biggest number. Teachers (and professors!) need to learn that the old stereotype of children as concrete thinkers is incorrect. This view is derived from an oversimplification of Piaget (who in fact said that young children are sometimes *too* abstract).[15] Although valuing his insights, contemporary research goes beyond Piaget to offer a nuanced and detailed account of mathematical development that stresses both unexpected competencies and misconceptions. The bottom line is that young children do not need to be prepared to learn math: They are already doing it and are quite capable of learning much more if they are properly taught.

Second, teachers need to rethink the content of early childhood math education, which is much more than counting or saying a few shape words or memorizing some number facts. Instead, early math education should be all about ideas, strategies, communication, and proof. The content is not simply addition; it is thinking and talking about addition to explain it. The content is strategies to solve addition problems and to check the answers. It is also having a positive and productive disposition to learning this kind of math as well as developing initiative, persistence, and focused engagement.

Third, teachers need to appreciate that the math children should learn is not trivial. We saw, for example, how the ideas underlying the counting words and answers to the question "How many?" are deep. Teachers cannot teach well if they do not understand what they are teaching, and they cannot assess well if they do not understand what the child should be learning to understand.

Fourth, children need active help in two major and related areas. The first is language. Children need to learn how to not only *do* such things as label shapes and count out loud but also *talk* about mathematical ideas and their own thinking. To talk in this way, children need to understand that math requires reasoning—to become aware of their own thinking, to believe that talking about it is useful and valued, and to appreciate the importance of communication and proof. Another way of conceptualizing this issue is to say that children need to learn to "mathematize"—that is, to make intuitions explicit and shareable, eventually in written form.

Fifth, teachers need to learn that intentional teaching is essential. Play is wonderful, but it is not enough, especially for helping children to talk about math and mathematize their intuitions. The National Association for the Education of Young Children, the major early childhood professional organization, in cooperation with the National Council of Teachers of Mathematics (2002) and the National Research Council (2009) endorse this position.

Sixth, teachers need to learn that good early childhood teaching seldom involves drill, narrow instruction, and textbooks. These methods do not work well for older children; the likelihood of success for young children is even smaller. Instead, a key element of teaching is to encourage active engagement by building on children's current level of understanding (remember the example of helping Ben to learn the tens numbers).

Seventh, to teach well, the teacher needs to learn to assess individual children's current level of understanding. As we saw, assessment and teaching are two sides of the same coin, and a key method of assessment is the clinical interview. Observation is important and may even occasion the start of an interview, but observation, similar to play, is not enough. Teachers should also understand that the interviews need not be as long and extensive as those shown here. Interviews can be integrated into classroom routines, such as circle time or book reading, and roving around the classroom, teachers can engage in short one-question interviews ("How did you do that?").[16]

CONCLUSION: THE TEACHER'S INTERMEDIARY INVENTIVE MIND

Teachers need to apply to the classroom all these ideas about children's mathematical development, the content of early math education, and the importance of intentional teaching and assessment, particularly the clinical interview. Professional development should help them make the application—that is, bridge the chasm between theory and practice.

William James, in his *Talks to Teachers* (1958), originally published in 1899, offers this still essential advice:

> You make a great, a very great mistake, if you think that psychology, being the science of the mind's laws, is something from which you can deduce definite programmes and schemes and methods of instruction for immediate schoolroom use. Psychology is a science, and teaching is an art; and sciences never generate arts directly out of themselves. An intermediary inventive mind must make the application, by using its originality. (pp. 23–24)

I take this to mean that the goal of professional development should be to help teachers understand the cognitive ideas (and the math, and the rest of the ideas already discussed) in such a way as to apply their understanding to the task of educating individual children in their classrooms. Teaching is an "art" in the sense that the teacher applies sophisticated knowledge to the individual case in an original, creative manner—artfully. Vague, general, and sentimental slogans about developmentally appropriate education or constructivism or children's developmental trajectories or the value of manipulatives and play are only minimally useful for the practical task of teaching. In fact, they may provide an excuse for not doing much of anything. The goal is to help teachers develop an intermediary mind that is sufficiently inventive and powerful to assess, interpret, and respond to individual children's minds as they learn specific content—in short, to use theory to guide meaningful and productive practice. How we can achieve this goal is the subject of subsequent chapters.

FOR REFLECTION AND ACTION

1. This chapter is organized around a series of video vignettes of one child, Ben, as he is interviewed across several years. What do you learn from viewing the video clips, both about mathematical development and about the use of clinical interviewing?

2. If you have worked directly with young children, you might think about similarities and differences between Ben's mathematical thinking and what you may have experienced. If there are differences, what might be some of the causes?

3. After reading this chapter and watching the video clips, might you consider using or adapting some of these video/interview techniques in designing and implementing professional development?

4. What difference might it make if teachers and others in early childhood education had a deeper understanding of children's mathematical thinking, as illustrated in this chapter?

5. In your experience, have you known an early childhood educator who embodied the goal of intentional teaching in mathematics and who was able to "apply to the classroom all these ideas about children's mathematical development, the content of early math education, and the importance of intentional teaching and assessment"? What kinds of experiences led to this level of comfort and competence?

CHAPTER 3 NOTES

1. I wish to acknowledge the useful comments and suggestions provided by Juanita Copley, Julie Diamond, Marilou Hyson, and Taniesha A. Woods.
2. Children in unschooled societies also do everyday math. It may take different forms in different cultures, but it is still everyday math, often of some complexity (Nunes, Schliemann, & Carraher, 1993).
3. Many thanks to Michael DeLeon.
4. It is interesting to think about children's interpretations of the clinical interview. What do they make of it? I often tell the children that I would like to learn about their thinking. But this probably makes little sense to young children (just as it does not to behaviorists). One first-grade child, Toby, whom you meet later and whom I interviewed for a good 45 minutes, said at the end, "Who are you?" Maybe children think that the strange interaction with the odd adult is a kind of game. But judge for yourself whether the somewhat antic interaction reveals insights into Ben's mathematical mind.
5. Many thanks to Azadeh Jamalian.
6. This is a good example of how the available evidence permits more than one plausible interpretation. Perhaps in this case, Ben simply thought that Ollie would prefer some bears to others. Janet did not ask him, so we do not have enough evidence to render a definitive judgment. The moral is this: Try to imagine alternative explanations and be modest about your own.
7. It is irrelevant that there were actually more than eight. Whatever the last number counted, right or wrong, he should believe that it indicates the total.
8. This immediate perception of the number is called *subitizing*, from the Latin word for "quickly" or "immediately." Young children can subitize small numbers of objects, such as two or three, and gradually expand their capacity to subitize, reaching a limit of about seven or so in adulthood.
9. Note that when talking about adding, I make putting-together or gathering motions with my arms. Gestures such as these are often associated with concepts and words such as "altogether," "take away," and many others.
10. Ben is not at all unusual. Zur and Gelman (2004) show that 3- and 4-year-olds generally can do the counting on that Ben did.
11. This is often called *metacognition* and is not easy for young children to do well (Pappas, Ginsburg, & Jiang, 2003).
12. Researchers need to learn this lesson, too. Current norms are based on current tasks. New tasks may reveal new norms. Why? The children did not change; the measurement procedures did.
13. This and subsequent quotations from this interview are in Ginsburg (1996, pp. 175–202).
14. This does not necessarily mean telling the children that they are wrong in seeing the equal sign as *makes*. I would argue that this view is legitimate in some circumstances and that the idea of the equal sign as equivalence is legitimate in others. For example, when you make three jumps on a number line and then two more jumps, you end up on five. *Ending up on* is essentially the same as *makes*, indicating that you do an operation to get a result or to end up somewhere. The important goal, I think, should be to teach children that both interpretations are legitimate and can be useful for different purposes. This is true in many

other areas of mathematics. For example, you can interpret 1/3 as a third of a pie or as 1 of 3 objects. These meanings are very different—one continuous and one discrete.

15. For a discussion of Piaget's views on this matter, see Piaget (1951) as well as Ginsburg and Opper (1988).

16. Socrates engaged in the peripatetic (literally walking around) method as he engaged his disciples in a kind of clinical interview discourse.

REFERENCES

Ginsburg, H.P. (1996). Toby's math. In R.J. Sternberg & T. Ben-Zeev (Eds.), *The nature of mathematical thinking* (pp. 175–202). Hillsdale, NJ: Lawrence Erlbaum Associates.

Ginsburg, H.P., & Opper, S. (1988). *Piaget's theory of intellectual development* (3rd ed.). Englewood Cliffs, NJ: Prentice-Hall.

James, W. (1958). *Talks to teachers on psychology: And to students on some of life's ideals.* New York, NY: W.W. Norton.

Miller, K.F., & Parades, D.R. (1996). On the shoulders of giants: Cultural tools and mathematical development. In R.J. Sternberg & T. Ben-Zeev (Eds.), *The nature of mathematical thinking* (pp. 83–117). Mahwah, NJ: Lawrence Erlbaum Associates.

National Association for the Education of Young Children & National Council of Teachers of Mathematics. (2002). *Early childhood mathematics: Promoting good beginnings.* Washington, DC: Author. Retrieved from http://www.naeyc.org/files/naeyc/file/positions/psmath.pdf

National Governors Association Center for Best Practices & Council of Chief State School Officers. (2010). *Common Core State Standards.* Washington, DC: Authors.

National Research Council. (2009). *Mathematics learning in early childhood: Paths toward excellence and equity.* Committee on Early Childhood Mathematics, C.T. Cross, T.A. Woods, & H. Schweingruber (Eds.). Center for Education, Division of Behavioral and Social Sciences and Education. Washington, DC: National Academies Press.

Nunes, T., Schliemann, A.D., & Carraher, D.W. (1993). *Street mathematics and school mathematics* (pp. 13–27). Cambridge, England: Cambridge University Press.

Pappas, S., Ginsburg, H.P., & Jiang, M. (2003). SES differences in young children's metacognition in the context of mathematical problem solving. *Cognitive Development, 18*(3), 431–450.

Piaget, J. (1951). *Play, dreams, and imitation in childhood.* (C. Gattegno & F.M. Hodgson, Trans.). New York, NY: W.W. Norton.

Zur, O., & Gelman, R. (2004). Young children can add and subtract by predicting and checking. *Early Childhood Research Quarterly, 19*(1), 121–137.

4

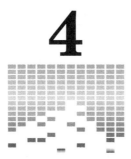

Goals for Early Childhood Mathematics Teachers

Juanita Copley

I have been privileged to facilitate mathematics professional development for current and prospective teachers of young children for the past 30 years. As part of my responsibilities as a university professor and grant administrator, I have worked in a variety of settings, both public and private, with preservice and in-service teachers. The Early Childhood Collaborative (Copley, 2001, 2004a) involved students preparing to be teachers and active teachers, some that had college degrees and others who had Child Development Associate credentials or associate's degrees. During those years, I have heard many goals expressed: some from prospective teachers ("I want to teach math so they understand . . . and not like I learned!"); some from experienced early childhood teachers ("I need to know what to do to teach the math that is important!"); some from university coordinators ("Our graduates need to be proficient in mathematics and experience a rigorous curriculum"); some from early childhood administrators or directors ("I want them to teach to meet our goals or competences"); some from parents ("I want my child to be ready for school and be good in math!"); and even some from children ("Can we do more 'fun' numbers?"). Although these goals are certainly important and interrelated, goals for early childhood teachers of mathematics encompass more coherence and focus.

To create effective professional development for teachers, it is essential for the goals to be clearly stated, understood, and agreed on by teachers, directors, and the professional developer. This type of goal setting is not an easy task, and one that develops over time. Let me illustrate with a recent example in Video 4.1: Director Explaining Changing Math Goals (see the About the Ancillary Materials page for details). Panda Path Early Learning Center is a public prekindergarten school in an urban, high-need area with a full-day program. Director Sara Hannes discusses her understanding of mathematics for young children, the growing mathematics focus of the Panda Path teachers, and the mathematics goals that have been developed over the past 3 years.

Many researchers and professional developers have proposed goals for teachers of young children. The National Board for Professional Teaching Standards (NBPTS; 2010) stated five general "propositions" that could be translated as goals: 1) Teachers are committed to students and their learning, 2) teachers know the subjects they teach and how to teach these subjects to students, 3) teachers are responsible for managing and monitoring student learning, 4) teachers think systematically about their practice and learn from experience, and 5) teachers are members of learning communities. Using the Early Childhood Longitudinal Study data, Palardy and Rumberger (2008) investigated the meaning of "a highly qualified teacher." They recommended that improving aspects of teaching such as instructional practices and teacher attitudes would be important goals and would provide a more robust association with student achievement. Leikin and Zazkis (2010) succinctly described teachers' goals as understanding what you know, what you need to know, and how that knowledge is learned and taught. Although all these goals could be classified as excellent, as stated, they are too general and theoretical to put into practice. As Ginsburg, Lee, and Boyd (2008) cautioned in a social policy report, teachers need to "avoid both vague theory and mindless practice" (p. 14). Without clear goals for early childhood teachers in mathematics, professional development could be unbalanced—that is, so theoretical that few or no practical skills could be applied or so practical that little or no thought could be used to reflect on practice or plan future instruction.

In this chapter, I discuss five specific goals for early childhood mathematics teachers, goals based on research as well as experiences of others and myself. This chapter is organized to discuss 1) a rationale and research supporting each goal, 2) recommendations based on experience and research, and 3) specific examples of the recommendations as implemented. This chapter reports many types of research relevant to the teaching of early childhood mathematics, and the resulting recommendations are based on my experiences as a professional developer.

The five professional goals in this chapter are not discrete and disconnected; however, I discuss each of them separately in order to clearly describe them and emphasize their uniqueness. Because of my work with both prospective and current teachers, I use the generic term *teacher* when I talk about goals that pertain to both groups; conversely, I use a specific identification if a goal is uniquely suited to either future teachers or current teachers. In addition, I state goals with high expectations, goals that need to be adapted to diverse professional development recipients and contexts. Although these goals are ambitious, they are important for all early childhood educators and, ultimately, for young children's learning in mathematics.The five goals to be discussed are as follows:

1. Know the content of math, including the math that should be taught to young children.

2. Understand how young children learn math and how best to teach them.

3. Know and implement appropriate teaching practices for effective mathematics learning.

4. Assess children's mathematics learning to inform instruction using formative in-class assessments.

5. Plan and organize to intentionally teach mathematics in a variety of settings.

GOAL ONE: KNOW THE CONTENT OF MATH

What is this mathematics content? Content knowledge has many forms. Math content knowledge can just be specific to mathematics, pedagogical content knowledge refers to understanding the content that is needed to teach the topic, and pedagogical knowledge is generally thought to be an understanding about how people learn and how best to teach them. For this first goal, I primarily focus on math content and pedagogical content knowledge; pedagogical knowledge is specifically discussed in the second goal.

Rationale and Research

Standards from all teaching associations list content-specific knowledge as a critical goal for teachers of young children (National Association for the Education of Young Children [NAEYC] & National Council of Teachers of Mathematics [NCTM], 2002; NBPTS, 2010; Lutton, 2012). The 2009 report by the National Research Council (NRC) specifically recommended that professional development for early childhood teachers should help them understand the mathematics necessary to teach young children, especially concentrating on 1) number, operations, and relations; 2) geometry, spatial relations, and measurements (with more mathematics learning time devoted to number than to other topic); as well as 3) the mathematical process goals (e.g., representing, problem solving, communicating).

Most researchers acknowledge that teacher knowledge is critical and suggest that the basic mathematical ideas that form the foundation for mathematics must be understood (Ball & Forzani, 2010; Lee, 2010; Scher & O'Reilly, 2009). However, the connections between the math you know and the math that must be taught are critical. Most professional developers propose that the math content emphasis should directly relate to the teachers' work, that the knowledge and the capacity to use it instructionally are important, and that interventions that include both content knowledge and pedagogical content knowledge have a larger positive impact on student learning than either content knowledge or pedagogical content knowledge alone (Clements & Sarama, 2008; Doerr, Goldsmith, & Lewis, 2010; Eisenhardt & Thomas, 2012; Scher & O'Reilly, 2009).

As further evidence of the importance of math content knowledge and pedagogical content knowledge, programs that focused mainly on teachers' instructional behaviors or "promising practices" demonstrated smaller influence on student learning than did programs whose content focused on teacher's knowledge of the subject connected to how that content can be taught (Carpenter, Fennema, Peterson, & Carey, 1998; Hill, 2010; Kennedy, 1998; Tassel et al., 2011; Tirosh, Tsamir, Levenson, & Tabach, 2011). Most importantly, the National Mathematics Advisory Panel (2008) found that direct assessments of teachers' actual mathematical knowledge—both content and pedagogical content—provided the strongest indication of a positive relation between teachers' content knowledge and their students' achievement.

Unfortunately, we know that the majority of early childhood teachers do not feel confident in or even express an active dislike in regard to mathematics content, including the mathematics that is found in the typical early elementary mathematics classroom (see Video 4.2: Teacher Describing Her Understanding of Math [go online to view the video]). This dislike can impede the learning and the teaching of mathematics content. Even more relevant to early childhood, female teachers' math anxiety and their ability to teach it resulted in more math anxiety and lower math achievement by female students (Beilock, Gunderson, Ramirez, & Levine, 2010). Since a majority of early childhood teachers are female, this is a reason for concern and provides an additional need to emphasize the goals of both math content and pedagogical content knowledge.

Recommendations for Professional Development

- *Introduce content through math problem-solving activities.* To learn the content of mathematics, teachers need to experience mathematics in a problem-solving context. Classwork, training sessions, or focused workshops on mathematics should focus on the foundational math content proposed by the NRC (2009) report (e.g., whole numbers, operations, geometry, and measurement) and connect to the higher levels of elementary mathematics. Problem-solving activities should be cognitively demanding and require teachers to think about their understanding (and perhaps experience frustration), to employ different solution strategies, and to engage in explicit discussions about the mathematics content. This learning can then be directly tied to the learning of young children so that they can experience and learn mathematics.

- *Make explicit connections between mathematics content and pedagogical content knowledge.* Early childhood teachers may know some of the foundational math concepts and procedures to teach young children. In my experience, what they do not know are the critical connections between 1) the math content and content needed to teach as well as 2) the foundational content and the later elementary mathematics content. Both of these connections need to be made explicitly. In some cases, I have begun with a problem that would typically be used in later grades and then connected the children's experiences with the foundational content in early childhood. In other cases, I model my use of pedagogical content knowledge as I teach a number operation lesson to young children and then debrief how that information relates to their own content knowledge about solving problems.

EXAMPLE: Experiencing a Cognitively Demanding Task to Learn Mathematics Content: Spirolaterals

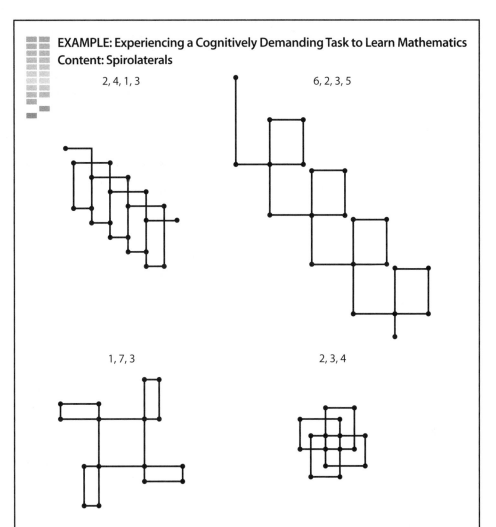

Figure 4.1. Spirolaterals with a three-digit pattern are closed. Those with a four-digit pattern are not closed.

Prekindergarten and kindergarten teachers in a 1-day focused workshop titled "Patterns and Algebra" worked in small groups to discover patterns in spirolaterals. Spirolaterals are designs created by a sequence of one-digit numbers on centimeter paper. To make a spirolateral, a specific procedure is used and repeated over and over again using the same sequence of one-digit numbers. Participants are encouraged to try a variety of examples and then to make some general statements about the patterns they discovered (see Figure 4.1). Typically a task for middle school students, this cognitively demanding task had many interesting outcomes. First, teachers (for the first time in many cases) understood the importance of patterns and why they teach pattern recognition, replication, and extension. Second, teachers eagerly worked with each other, and because of the detailed procedures necessary to the task, they realized how important it was to work together, question each other, and analyze their creations. Third, teachers discovered that they "enjoy this kind of math,"

(continued)

> (*continued*)
> and when they were told they were going through the algebraic process when writing their generalizations, they were excited! In fact, a majority of the 300 participants who took this workshop listed this activity as their favorite and the "most important to their understanding." And finally, teachers readily connected their learning regarding patterns to their students' understanding of patterns and remarked about the importance of children's persistence, interactions with their peers while solving problems, and tasks that are engaging.

GOAL TWO: UNDERSTAND HOW CHILDREN LEARN MATHEMATICS

Although early childhood teaching standards often include knowledge of child development as a goal for teachers, they typically mention it as an understanding of the "whole child" (NBPTS, 2010; NAEYC & NCTM, 2002). The position statement by the National Association for the Education of Young Children (NAEYC) and the National Council of Teachers of Mathematics (NCTM) states that teachers should "base mathematics curriculum and teaching practices on knowledge of young children's cognitive, linguistic, physical, and social emotional development" (2002). Yes, understanding the "whole child" is important and valuable; however, a general understanding of cognitive development is not sufficient when we think about how young children learn mathematics. In fact, I believe that you really do not understand the "whole child" if you do not know how they learn math. Thus the focus of this goal is specific to mathematics and how children understand the content of mathematics.

Rationale and Research

Many researchers emphasize the goal of understanding children's mathematical thinking as one that is essential to teaching (Philipp et al., 2007; Sarama,

> **EXAMPLE: Connecting Content Knowledge with Pedagogical Content Knowledge: Precise Geometric Vocabulary**
>
> During a mathematics class for prospective teachers, participants were involved in an upper elementary geometry activity that required them to create shapes that would fit in an appropriate section on a Venn diagram. To do this, they had to understand the defining attributes of a variety of shapes (e.g., rhombus, quadrilateral, rectangle). The activity required a great deal of interaction, clarification of precise geometry vocabulary, and discussion of misconceptions (e.g., defining incorrectly that a rectangle has two long sides and two short sides). After their struggles with math content knowledge in geometry, the prospective teachers were excited to learn how best to teach the defining attributes of shapes. They were eager to do the center activities that required them to post shapes on two-column posters titled "Rectangles" and "Not Rectangles" and "Rhombi" and "Not Rhombi." They were especially interested in ways to use Venn diagrams as a sorting tool that young children can understand.

2002; Sarama & Clements, 2009; Saracho & Spodek, 2008). Understanding how children learn math is not a simple task. Recently, there has been an abundance of research involving the mathematical understanding of young children in the foundational areas of number, operations, and geometry. Professional developers, researchers, university professors, and early childhood directors advocate that this goal includes 1) introducing recent research regarding children's mathematical thinking early in teachers' education, 2) attending to "teaching–learning paths" that match teaching suggestions with learning progressions in the foundational math content areas, and 3) observing young children on video or working directly with them at the same time as teachers are instructed on how children learn math (Docrr et al., 2010; Ginsburg, Lee, & Boyd, 2008; Kennedy, 1998; NRC, 2009).

A recent study that involved a group of early childhood teacher-leaders emphasized the importance of understanding how children learn math. The teacher-leaders were involved in 6 days of workshops on number, geometry, and measurement topics that focused on math content knowledge, math pedagogical knowledge, and developmental trajectories or paths (Clements & Sarama, 2009). They reported that from their perspective, understanding the developmental progressions had the highest potential for promoting effective learning by both children and teachers in early childhood settings. The teacher-leaders also reported that the linking of the learning paths to the teaching strategies helped them identify some benchmarks and, more important, helped them know what to teach and when to teach it (Wilson, 2009).

Most of this goal has emphasized the learning paths and what teaching–learning progressions tell the teacher about how children learn math (see Video 4.3: Teacher Describing His Instructional Games [go online to view the

---

EXAMPLE: Connecting Teaching–Learning Paths to Math Content Knowledge and State Mathematics Standards

Six-day workshops conducted for prekindergarten teachers in four states followed a particular sequence of events each day. A different topic was discussed each day (e.g., beginning number concepts, geometry, number operations, measurement). During each workshop, participants learned 1) some mathematics that was specifically relevant to their teaching, 2) teaching–learning paths that related to the topic, and 3) the connection of the teaching–learning paths to the state's or the organizations' math standards. In many cases, specific teaching–learning paths did not relate directly or indirectly to the state's or organization's standards. For example, in one state, there were no objectives, goals, or standards for comparing number quantities. In another state, there were no objectives, goals, or standards for composing or decomposing shapes; the only geometry standards dealt with naming shapes and position words. This was an important discovery for the participants. The teachers acknowledged the importance of using the teaching–learning paths to help them identify important concepts, teach them, and regardless of the lack of objectives or standards, plan their curriculum around the skills and concepts identified by the teaching–learning paths.

> **EXAMPLE: Understanding Children's Thinking Using Clinical Interviews: Ginsburg's (1997)** *Entering the Child's Mind: The Clinical Interview in Psychological Research and Practice*
>
> Graduate students in early childhood education were required to conduct clinical interviews with young children. Using the procedures for clinical interviews outlined in the Ginsburg text, students described "windows" into children's thinking in mathematics. As part of their work in the class, they also used a variety of other instruments, tests, checklists, and general observations. Students reported (and the instructor noted) that the clinical interviews revealed more information about a child's thinking and how they learned than any other method or instrument.

video]). It is also important to note that mathematics is present in much of children's spontaneous play and everyday activities (see Video 4.4: Two Girls Playing a Game [go online to view the video]; Ginsburg, Inoue, & Seo, 1999; Saracho & Spodek, 2008; Seo & Ginsburg, 2004). Although play has the potential of involving mathematics learning, that potential depends largely on the teachers' ability to appropriately facilitate both the play and the learning. Dockett and Perry (2010) suggested that when educators have 1) mathematics content knowledge, 2) a pedagogical repertoire that includes play, and 3) an understanding of the connections among these, then young children's understanding and dispositions for mathematics can be increased.

Recommendations for Professional Development

- *Embed understanding of specific mathematical learning progressions or paths within pedagogical content knowledge.* After I teach math lessons in early childhood classrooms, I require teachers to reflect on what they observed about how children were learning during the instruction. These types of discussions help me introduce specific learning progressions as they pertain to their observations, and it helps teachers connect the progressions to their own teaching. (See Chapter 3 for an excellent example of embedding learning progressions with pedagogical content knowledge.)

- *Provide opportunities for teachers to listen and observe children intentionally.* Teachers need opportunities to interview children so that they better understand how young children learn mathematics. The use of children's work samples, observations of children at play, diagnostic interviews, and video clips of young children as they do mathematics help teachers observe intentionally.

    Video 4.5: Young Child Making a Block Construction (from an hour-long tape; go online to view the video) shows a young boy solving his own block construction problem provides an opportunity for discussion about persistence and the joy of problem solving even at a young age.

- *Pair experienced early childhood teachers with experienced mathematics teachers.* Mathematics curricula and guidelines are often written by educators who have little or no experience with teachers of young children. Similarly, early childhood curricula and guidelines are sometimes written by experts in early childhood and literacy, socioemotional development, or pedagogy that have excluded mathematics. By pairing expert teachers with these different kinds of experiences, an understanding of mathematics and young children's development and the relationship between the two can be investigated with promising outcomes in regard to curricular and instructional interventions.

GOAL THREE: KNOW AND IMPLEMENT APPROPRIATE TEACHING PRACTICES FOR EFFECTIVE MATHEMATICS LEARNING

The goal of knowing and implementing appropriate practices for mathematics teaching has likely received the most attention in education, seminars, and professional development. In the past, educators may have skipped over content knowledge and an understanding of children and focused instead on pedagogy, which is an ineffective practice!

Rationale and Research

Certainly, appropriate teaching strategies should be used if children are to learn mathematics. Ball and Forzani state, "Although teachers need to thoroughly understand the material they teach, that is not the same as knowing how to teach it" (2010, p. 10). Researchers agree that all teachers need to be exposed to appropriate pedagogy and learn to think critically about their teaching practices (Bogard, Traylor, & Takanishi, 2008; Boonen, Kolkman, & Kroesbergen, 2011; Ginsburg et al., 2008; Palardy & Rumberger, 2008). The problem is that early childhood educators currently teach so little math, and what they do teach is often done poorly (NRC, 2009).

So what are the most effective practices to teach mathematics to young children? Most researchers and educators agree on a large list of some general practices, ones that are applicable in various forms to all children: teaching for conceptual understanding, choosing and designing rich tasks, using mathematical tools, creating multiple representations, facilitating discourse, interacting with students, analyzing students' errors, attending to and using math language, teaching for problem solving, designing intentional lessons, introducing sensitivity to patterns, encouraging a willingness to experiment, responding to students' thinking, posing mathematical questions, and explaining procedures and ideas (Cadima, Leal, & Burchinal, 2010; Merritt, Rimm-Kaufman, Berry, Walowiak, & McCracken, 2010; Saracho & Spodek, 2008; Thames & Ball, 2010). Obviously, the list of practices in this paragraph could be expanded to a book, and a very large book at that! For the purposes of this chapter, I simply highlight research on a few and then include the highlighted ones within the recommendations.

High-quality interactions between teachers and children as they learn mathematics is a general practice that has been strongly supported by research. These quality interactions can be described succinctly as the ability to communicate the math content in a meaningful manner. But what does that mean?

High-quality interactions in mathematics involve 1) the mathematization of children's experiences by redescribing what they have experienced in math language, 2) engaging students with particular math questions about subjects that interest them, 3) questioning and prompting to encourage children's thinking, 4) reflective listening to children's ideas about math, 5) scaffolding or using "assisted performance" comments to support their problem solving, 6) making statements that summarize a particular set of math explorations, and 7) encouraging children to communicate with their peers or to engage other sources (e.g., technology) to solve math problems (Boonen et al., 2011; Hamre et al., 2012; NRC, 2009).

The use of manipulatives or mathematical tools has been widely advocated by professional developers and teachers as a practice that is necessary to communicate math content in a meaningful way. Unfortunately, although teachers believed that hands-on activities are of great importance, they are not able to analyze the specific manipulative or the way it could be used most effectively. Teachers received manipulatives in many early childhood math projects or grants, but they rarely had ideas about how to teach with them. Instead, they allowed students to use them for discovery and invention (which is good), but they did not know how to use them intentionally to facilitate mathematics learning (Bussi, 2011; Copley, 2004a, 2004b; Thorton, Crim, & Hawkins, 2009).

Does professional development in effective practices work? Generally, the answer is yes, *when* it is specifically connected to a curriculum, involves an understanding of how children learn, and contains a cycle of assessment of learners, active teaching, observations, and reflection on the specific practices taught and implemented. Specifically, professional development projects that combined practice-focused workshop sessions with classroom coaching resulted in an increase in effective practices, the use of math-mediated language with the greatest increase occurring during the coaching phase of the treatment (Eisenhardt & Thomas, 2012; Fukkink & Lont, 2007; Rudd, Lambert, Satterwhite, & Smith, 2009; Sarama & Clements, 2009).

Recommendations for Professional Development

- *Provide models focusing on child–teacher interactions using technology or face-to-face models.* Language and child–teacher interactions involve

---

EXAMPLE: Editing Teaching Tapes and Sharing the Highlights: C3 Peer Coaching (Copley, Hawkins, Houston, & Padron, 2003–2007)

Prekindergarten teachers in a 3-year coaching grant for recommended practices in mathematics and literacy made five video clips of their recommended practices. They then had to work with a group of their peers, edit the clips to show the "best of the best," and include their reflections and observations. The resulting DVDs were then shared with other grant participants. Teachers reported that this aspect of the grant made the greatest difference in their teaching practices.

questioning strategies; responding to children's actions, questions, or work; facilitating children's oral language; and introducing mathematical vocabulary in a meaningful way. By providing models that focus on these interactions, teachers may be able to increase their use of effective practices and hopefully increase mathematics learning. The use of technology to film expert teachers as they model lessons, the use of web-based examples, or modeling by in-class coaches could provide some models for these practices.

- *Encourage classroom observations of practices and include opportunities for reflections and feedback.* Teachers need to practice their use of research-based practices. They need to spend time watching themselves (using video clips) or listening to the evaluations of others as they interact with their students, ask questions, respond to their needs, and prompt their thinking. It is essential that there be opportunities for reflection on their instruction and that they be given feedback by a significant other (e.g., a coach, a peer, an expert).

- *Teach specific practices that match a curriculum along with specific mathematics tools.* The curriculum and the practices designed to deliver the curriculum should be taught together. Often educators teach isolated tasks or activities that they consider cute or fun with little or no connection to math content or how children learn math. Similarly, educators should be taught the purposes of specific manipulatives, learn how to use them with children effectively, and identify mathematical concepts that can be taught using the tools (see Video 4.6: Teacher Describing Her Use of Manipulatives [go online to view the video]).

GOAL FOUR: ASSESS APPROPRIATELY TO INFORM INSTRUCTION USING FORMATIVE, IN-CLASS ASSESSMENTS

Assessment of mathematics achievement is a matter of national interest and concern in schools in the United States. Typically, assessment for the purpose of evaluation begins in third grade and continues throughout a child's academic career. However, if we believe that mathematics is foundational for a young child, it is important that the assessment of young children's learning begins early with another purpose, that of informing instruction.

Rationale and Research

Good classroom-based formative assessment gives teachers the chance to make changes in instruction at the point of children's needs and can facilitate their mathematics learning. To inform instruction, teachers need to be able to use in-class assessments, interviews, performance tasks, and continuous observations; these snapshots represent various stages of children's development and learning. Teacher educators and professional standards recommend that early childhood teachers should understand and be able to implement a variety of formative assessment strategies (Bogard, Traylor, & Takanishi, 2008; Ginsburg & Dolan, 2011; Ginsburg et al., 2006; NAEYC & National Association of Early Childhood Specialists in State Departments of Education [NAECS/SDE], 2003; NRC, 2009; Saracho & Spodek, 2008).

Unfortunately, a review of teacher education and training objectives clearly illustrates that early childhood teachers have not been educated to assess young children's development and learning in mathematics. Many researchers question if teachers have the available knowledge to assess what their students know and do not know about mathematics concepts. In general, they have found that teachers make assessments based on their general impressions of children's abilities in order to communicate to parents or complete report cards, rather than to inform instruction and specifically address misconceptions, inaccuracy of skills, or advanced reasoning.

Kilday, Kinzie, Mashburn, and Whittaker (2012) examined the associations between teachers' judgments of children's math skills using a skill rating scale (one to five, with one indicating that a child has not demonstrated skill to a five indicating that the child is proficient in a skill) and two standardized tests (TEMA-3 and M-TEAM) to identify specific math skills in number, geometry, and measurement. The results indicated that teachers can accurately determine whether students are above or below average in general mathematics proficiencies but that they cannot appropriately rate students on specific mathematics skills. To inform instruction and address student needs and progress, teachers need to know more than a child's general math ability.

Prusaczyk and Baker (2011) reported the results of a project that was based on Cognitively Guided Instruction (CGI; Carpenter, Fennema, Franke, Levi, & Empson, 1999), a curriculum that closely ties assessment and instruction using a framework of mathematics word problem categories. A distinct assessment plan was presented to teachers in the project: 1) practice identifying achievement levels using the CGI framework and adjusting instruction, 2) learn how to assess student work using rubrics, and 3) meet four times in grade-level clusters

---

EXAMPLE: Assessing Skills and Understanding Through Games: The Snake Game

Children play The Snake Game, a South American game that requires young children to completely fill up a snake board made up of 25 sections. To move, they guess how many counters are hiding in their partner's hand (zero to three). If the guess is correct, the guesser fills up the spaces with that number of counters. If the guess is incorrect, the partner fills up the spaces on his or her snake. After teachers play the game with their students, they meet together and outline what skills/concepts they could assess by watching children play the games. In this scenario, they decided they could assess if children 1) could subitize quantities of 2 or 3, 2) could match counters with spaces using one-to-one correspondence, and 3) knew the meaning of zero. They also decided that they needed to ask some questions during the game to clarify the child's thinking. In fact, they became strong advocates not only for watching children play a game; they also wanted to be players! With this one model, teachers selected children's favorite games and generated a list of what they could assess with the games. They responded favorably to the use of games as an assessment tool.

> **EXAMPLE:** *Not* **Asking the Right Questions and Probing Children's Responses: Counting to 100**
>
> Recently, a parent received a note from the prekindergarten teacher. Her son had performed very well at the beginning of the year by counting to 100. Then, at the end of the year, he could only count to 20. When the concerned parent asked the child about what happened, he responded, "I am just tired of counting to 100 ... I could even count to 200 cause numbers just keep going but I don't want to! So I told the teacher I could only remember to 20!" The teacher's assessment was clearly incomplete! She should have asked the same question the parent did.

to assess children's abilities as described in the state learning standards for mathematics. The researchers reported that teachers improved in their assessment of student learning as defined by the CGI framework, and they attributed the results to the three training strategies that "encouraged teachers to see the power of formative assessments that can enhance student learning" (p. 115).

Recommendations for Professional Development

- *Teach appropriate assessment methods specific to mathematics.* Teachers need experience with a variety of assessment strategies and support using interviews, performance assessments, checklists, observations of games that address number skills, some math-specific questions to ask at center time, and other ideas that match the curriculum and the teaching–learning paths. In most cases, this is new learning for both experienced and inexperienced teachers, so practice, expert support, reflective discussions, and collaborative work with their peers would be necessary.

- *Illustrate the use of open-ended questions, response strategies, and additional probes.* Teachers need to learn methods to encourage children to use their words or actions to show what they know. Interactions with students can provide assessment opportunities if teachers know how to ask questions, probe for additional information, and respond appropriately. This type of assessment must be built into teaching and should be strongly connected to the curriculum that is used.

GOAL FIVE: INTENTIONALLY PLAN AND ORGANIZE TO TEACH MATHEMATICS IN A VARIETY OF SETTINGS

I frequently ask teachers to tell me when they teach mathematics. In almost every case, they tell me that they teach math all day long and that "math is everywhere!" When I ask them to explain what they mean, they say statements such as, "We count everything wherever we go," or "We walk to the buses in a pattern," or "We sing the shape song as a beginning routine almost every day," or "We read a math story once a week." When I inquire about focused math time when children are actively learning mathematics, they remind me that

> **EXAMPLE: "Where's the Math?" Workshop: Teachers Evaluating Their Classrooms Using an Environmental and Lesson Checklist**
>
> During a 1-day workshop, prekindergarten teachers were introduced to mathematical practices from an early childhood perspective along with the state guidelines or standards for early childhood. They viewed video clips and lesson plans, watched as they were implemented, and then participated in center activities, read-aloud activities, routines, transitions, circle time, play, games, and small-group lessons. Participants reflected on where they saw math intentionally taught. At the end of the day, they received an environmental and lesson checklist. They then returned to their site and used the checklist to find "where the math is" in their day and classroom environment. The results of this first checklist have proved to be an excellent reflective tool and one that teachers use again after a few months with excellent improvements.

they must spend most of their time on literacy, that they have a half-day program, or that they really do not have time to do math every day.

Rationale and Research

These casual, unfocused responses concern me for several reasons. Research strongly advocates for 1) the intentional teaching of mathematics in early childhood; 2) the need for direct, formal development of children's concepts in mathematics and explicit plans for mathematics (i.e., a curriculum); 3) a variety of instructional methods and settings (e.g., routines, whole group, small group, play, routines, transitions), including teachable moments as they occur; and 4) a connection between family and community programs in which math learning can be promoted and learned (NAEYC & NCTM, 2002; NRC, 2009).

Let us talk about a few of these settings. First, and most important, are the individual classrooms. The teacher must plan intentionally to teach mathematics every day using a curriculum that includes the formal development of children's concepts in mathematics together with an instructional focus on mathematics and connections to the teaching–learning paths. In most classrooms, this is not happening (see Chapter 2); however, it should be a goal for every early childhood teacher. To best meet the needs of every child (those who are disadvantaged as well as those who need and want to be intellectually challenged), teachers need to teach mathematics intentionally, using a well-designed curriculum connected to the teaching–learning paths; changing instruction as needed using formative, in-class assessments; and being aware of interventions and using them when needed. In his position article, Torff (2011) addresses the importance of teachers' beliefs and curriculum: "A rigor gap emerges in which disadvantaged students are judged to require less rigorous curriculum than that afforded their more privileged peers" (p. 22). Saracho and Spodek (2008) resisted this idea and cautioned that teachers should not place limits on young children's mathematical thinking. The National Mathematics Advisory Panel (2008) recommended that the long-term effects of early interventions that strengthen mathematical knowledge should be researched "urgently" with a particular focus on at-risk learners and that teachers should be trained in the specific intervention. Unfortunately, they make no mention of the young learner who needs to be challenged.

Mathematics should not just be intentionally taught during the focused math time (see Video 4.7: Teacher Describing Using Math Throughout the Day [go online to view the video]). Indeed, a combination of teaching strategies and planned experiences can be generated to promote math learning in centers (e.g., block or sand center, game center, technology center, play store, flower shop, aerobics, or hospital centers), routines (e.g., songs, storytime, movement activities), transitions (e.g., lining up in groups of 5 or 10, comparing quantities by matching in a 2-by-2 line, or using sound patterns or beats), project work that involves science or literacy concepts but primarily focuses on mathematics, or whole group math games or counting experiences. Researchers agree that *if* instruction is planned and organized, these experiences can promote children's mathematics learning, especially if they are included throughout the day (NAEYC & NCTM, 2002; Saracho & Spodek, 2008).

Intentionally planning for mathematics instruction can also involve the whole school or center, or even the school district. Teacher specialists (e.g., behavioral interventionists, special education teachers, music teachers, physical education teachers) can and should also be involved in planning intentionally for mathematics (see Video 4.8: Director Explaining Intentional, Appropriate, and Playful Math [go online to view the video]). In addition, teachers can work together developing experiences/activities that provide opportunities for children to learn math, share materials, and provide additional input on specific assessment items. When teachers belong to a professional learning community and plan together with a team of their peers, they are provided with valuable

---

EXAMPLE: Implementing Curriculum (*Freda Frogonaut and the Lily Pad Space Station: Mentoring Mathematicians*) and Sharing Work Samples Outcomes in a Variety of Classroom Settings (Gavin, Copley, Sheffield, Chapin, & Casa, 2010)

An advanced curriculum project for kindergarten students in geometry and measurement was created and implemented in three different states in high-need, low-performing settings. Funded by the National Science Foundation, this project required students to write and communicate orally with Freda, a frogonaut on the Lily Pad Space Station. Initially, kindergarten teachers involved in this project were quite doubtful that their children could actually do and understand the mathematics in this unit. Grounded in the project-based approach as well as a philosophy of "writing to learn," the researchers were also unsure of the results of this advanced curriculum with the many writing samples required to communicate children's thoughts, reasoning, and solutions. When the project was finished, both researchers and teachers were all amazed at the children's work samples, their ability to communicate their reasoning orally, their written work, and their achievement levels. The analysis of work samples from all settings especially contributed to a positive belief that all children can learn mathematics!

> **EXAMPLE: Creating and Attending Family Math Nights for Early Childhood Families**
>
> In a large urban school (i.e., more than 1000 students in pre-K to Grade 5) in a low socioeconomic status (SES) area, the first parent meeting for prekindergarten and kindergarten was attended by only 10% of the families. To communicate with families about the importance of mathematics learning to their young children, a Family Math Night was planned. Attendance was advertised as a "date night," meaning that children had to attend with an adult and an adult had to attend with their young children. Stations were set up in the gym where parents and children could play math games, enter estimation contests, and participate in a variety of activities together. The value of mathematics learning was illustrated and rewarded with math-related door prizes donated by community organizations. By the end of the year, more than 80% of the families attended the monthly Family Math nights, and excitement about mathematics became a community and family event.

support and given opportunities to learn. The mathematics outcomes of the students that belong to these classrooms have been quite positive for both students and teachers (Copley, in press; Heitin, 2011).

Finally, when teachers intentionally organize and plan to teach mathematics, they should think about the families and the community that surround the school. Although there is little evidence that specific programs can enhance mathematics learning, partnerships between families and members of the community do have the potential to increase the mathematical learning of young children. The media and technology (e.g., educational television, computer software) and community programs (e.g., museums, exhibits, art shows) can all be places where teachers can organize and intentionally teach mathematics. Perhaps most important, informing parents about the importance of mathematics, along with some suggested activities and games, has demonstrated promising outcomes for young children (Coates & Thompson, 1999; NRC, 2009).

Recommendations for Professional Development

- *Plan for focused, intentional math experiences so that mathematics can be actively introduced.* From the data, it appears that prekindergarten teachers do not purposefully plan math experiences. Instead, they do math when they have time, include it with other content, or incorporate it as activities with the calendar. This recommendation, if implemented and combined with the first three goals, would make a difference in young children's learning.
- *Organize for instruction by employing a variety of settings (i.e., whole group, routines, small groups, centers, partner activities, outside activities).* Just as books and writing instruments can be used in every center, so can math materials and representations. Just as children enjoy listening to a story, teachers can use stories to introduce and act out word problems. Just as teachers meet students in small groups to introduce or review

skills or concepts in language or phonemic awareness, they can meet students in small groups to review the quantity of five or other concepts. Just as children can do motor activities during transitions, they can also combine the activities with counting or measuring concepts.

- *Consider curricula and/or other projects that invite teachers to share ideas across classrooms or settings.* Encourage the establishment of professional learning communities, teacher pairings, or school sharing groups to provide opportunities for more mathematics learning by both students and teachers.

- *Generate programs with possible connections among mathematics and the families and community in teachers' schools or centers.* Use online technology to investigate the community resources that can promote mathematics learning. Encourage school family nights that focus on the importance of mathematics and appropriate at-home activities.

CONCLUSION

Five goals for teachers were described in this chapter. From the perspective of a beginning teacher or a teacher who has never focused on mathematics, there are too many goals, and they are overwhelming in scope and content. From the perspective of an experienced mathematics teacher, teacher educator, or professional developer, they are not described in enough depth and specificity, and many questions are left unanswered. I understand both perspectives.

As a teacher educator, I have worked with mathematics and early childhood teachers for more than 30 years. My goals for teachers have become more focused (i.e., important mathematics should be taught intentionally every day in early childhood settings), more conceptually and developmentally based (i.e., teachers need to teach the concepts of math with knowledge of how young children learn mathematics), and more flexible (i.e., teachers learn differently and a variety of strategies for professional development must be employed). If assimilated, the goals I have described in this chapter could develop early childhood teachers professionally, improve their practice, and most important, increase young children's mathematics achievement and enjoyment of math learning. From my perspective, the overall goal is for young children to

---

EXAMPLE: Planting, Growing, and Selling Produce in a Farmers' Market: A Community and School Project in Prekindergarten

An early childhood graduate student teaching in an urban, low-SES prekindergarten class created a math–science project in his school. His young students planted vegetable seeds in large cement containers inside the school courtyard. They watered and measured their plants as they grew and then harvested them with great success. When they took them to the local farmers' market, they made more than $300 in profit! The result: mathematics *and* science *and* community involvement!

experience, learn, and understand mathematics that provides a strong foundation for lifelong learning.

FOR REFLECTION AND ACTION

1. Copley begins this chapter with a video example of an early childhood director describing her goals and her center's goals for mathematics education and how these have evolved over time. If you are a professional development designer or provider, it might be valuable to have similar conversations with teachers in your area.
2. The chapter identifies five important goals for those who aim to help young children learn about and become interested in mathematics. In your experience, which of these goals is the most challenging for teachers to achieve? Why might that be the case?
3. Copley offers specific professional development recommendations intended to strengthen teachers' competence in each of these five goal areas. If possible, try to implement several of these recommendations in your work as a professional development planner or provider.
4. Copley describes the characteristics of "high-quality interactions in mathematics." The downloadable content that accompanies this book (as well as the links in the e-book version) includes many video examples of such interactions between adults and young children. You might see if you can identify examples of each of these characteristics and consider how to help early childhood educators implement them more frequently.
5. The chapter concludes with the author's reflections on how her own goals for teachers' professional development in mathematics have changed over the years. Have your goals changed over time? Have they perhaps changed or been clarified as a result of reading this chapter?

REFERENCES

Ball, D.L., & Forzani, F.M. (2010). What does it take to make a teacher? *Kappan, 92*(2), 8–12.
Beilock, S.L., Gunderson, E.A., Ramirez, G., & Levine, S.C. (2010). Female teachers' math anxiety affects girls' math achievement. *Proceedings of the National Academy of Sciences, 107*(5), 1860–1863.
Bogard, K., Traylor, F., & Takanishi, R. (2008). Teacher education and PK outcomes: Are we asking the right questions? *Early Childhood Research Quarterly, 23*(1), 1–6.
Boonen, A.J.H., Kolkman, M.E., & Kroesbergen, E.H. (2011). The relation between teachers' math talk and the acquisition of number sense within kindergarten classrooms. *Journal of School Psychology, 49*(3), 281–299.
Bussi, M.G.B. (2011). Artefacts and utilization schemes in mathematics teacher education: Place value in early childhood education. *Journal of Mathematics Teacher Education, 14*(2), 93–112.
Cadima, J., Leal T., & Burchinal, M. (2010). The quality of teacher-student interactions: Associations with first graders' academic and behavioral outcomes. *Journal of School Psychology, 48*(6), 457–482.
Carpenter, T.P., Fennema, E., Franke, M., Levi, L., & Empson, S. (1999). *Children's mathematics: Cognitively Guided Instruction.* Portsmouth, NH: Heinemann.

Carpenter, T.P., Fennema, E., Peterson, P.L., & Carey, D.A. (1988). Teachers' pedagogical content knowledge of students' problem solving in elementary arithmetic. *Journal for Research in Mathematics Education, 19*(5), 385–401.

Clements, D.H., & Sarama, J. (2008). Experimental evaluation of the effects of a research-based preschool mathematics curriculum. *American Educational Research Journal, 45*(2), 443–494.

Clements, D.H., & Sarama, J. (2009). *Learning and teaching early math: The learning trajectories approach.* New York, NY: Routledge.

Coates, G.D., & Thompson, V. (1999). Involving parents of four- and five-year-olds in their children's mathematics education: The FAMILY MATH experience. In J.V. Copley (Ed.), *Mathematics in the early years* (pp. 205–211). Reston, VA: National Council of Teachers of Mathematics.

Copley, J.V. (2001). The early childhood collaborative: Communities of discourse. *Teaching children mathematics, 8*(2), 100–103.

Copley, J.V. (2004a). The early childhood collaborative: A professional development model to communicate and implement the standards. In D. Clements & J. Sarama (Eds.), *Engaging young children in mathematics: Standards for early childhood mathematics education* (pp. 401–414). Mahwah, NJ: Lawrence Erlbaum Associates.

Copley, J.V. (2004b). Preparing teachers to provide challenging mathematics to young children. In *Challenging Young Children Mathematically* (pp. 71–80). National Council of Supervisors of Mathematics (NCSM) Monograph Series, vol. 2. New York, NY: Houghton Mifflin, Leadership in NCSM Mathematics Education.

Copley, J.V. (in press). *Nita's preK playbook.* Boston, MA: Pearson's Learning Solutions.

Copley, J.V., Hawkins, J., Houston, R., & Padron, Y. (2003–2007). *C3 coaching grant for early childhood educators.* Washington, DC: Department of Education.

Dockett, S., & Perry, B. (2010). What makes mathematics play? In L. Sparrow, B. Kissane, & C. Hurst (Eds.), *Shaping the future of mathematics education: Proceedings of the 33rd Annual Conference of the Mathematics Education Research Group of Australasia* (pp. 715–718). Fremantle, Australia: Mathematics Education Research Group of Australasia.

Doerr, H.M., Goldsmith, L.T., & Lewis, C.C. (2010). *Mathematics professional development: Research brief.* Reston, VA: National Council of Teachers of Mathematics.

Eisenhardt, S., & Thomas, J. (2012). Early numeracy intervention: One state's response to improving mathematics achievement. *NCSM Journal of Mathematics Education Leadership, 14*(1), 28–36.

Fukkink, R., & Lont, A. (2007). Does training matter? A meta-analysis and review of caregiver training studies. *Early Childhood Research Quarterly, 22*(3), 294–311.

Gavin, K., Copley, J., Sheffield, L., Chapin, S., & Casa, T. (2010). *Freda frogonaut and the lily pad space station.* Unpublished report for M2 Mentoring Mathematicians, funded by the National Science Foundation.

Ginsburg, H.P. (1997). *Entering the child's mind: The clinical interview in psychological research and practice.* New York, NY: Cambridge University Press.

Ginsburg, H.P., & Dolan, A.O. (2011). Assessment. In F. Fennell (Ed.). *Achieving fluency: Special education and mathematics* (pp. 85–103). Reston, VA: National Council of Teachers of Mathematics.

Ginsburg, H.P., Inoue, N., & Seo, K. (1999). Young children doing mathematics: Observations of everyday activities. In J. Copley (Ed.), *Mathematics in the early years* (pp. 88–100). Reston, VA: National Council of Teachers of Mathematics.

Ginsburg, H.P., Lee, J.S., & Boyd, J.S. (2008). Mathematics education for young children: What it is and how to promote it. *Society for Research in Child Development Social Policy Report, 22*(1). Retrieved from http://files.eric.ed.gov/fulltext/ED521700.pdf

Hamre, B., Pianta, R.C., Burchinal, M., Field, S., LoCasale-Crouch, J., Downer, J.T., ... Scott-Little, C. (2012). A course on effective teacher-child interactions: Effects on teacher beliefs, knowledge, and observed practice. *American Educational Research Journal, 49*(1), 88–123.

Heitin, L. (2011). Pairing up. *Education Week: Teacher Professional Development Sourcebook, 5*(1), 26. Retrieved from http://www.edweek.org/tsb/articles/2011/10/13/01coteach.h05.html

Hill, H. (2010). The nature and predictors of elementary teachers' mathematical knowledge for teaching. *Journal for Research in Mathematics Education, 41*(5), 513–545.

Kennedy, M.M. (1998). *Form and substance in in-service teacher education* (Research Monograph No. 13). Arlington, VA: National Science Foundation.

Kilday, C., Kinzie, M.B., Mashburn, A.J., & Whittaker, J.V. (2012). Accuracy of teacher judgments of preschoolers' math skills. *Journal of Psychoeducational Assessment, 30*(2), 48–158.

Lee, J. (2010). Exploring kindergarten teachers' pedagogical content knowledge of mathematics. *International Journal of Early Childhood, 42*(1), 27–41.

Leikin, R., & Zazkis, R. (2010). Teachers' opportunities to learn mathematics through teaching. In R. Leikin & R. Zazkis (Eds.), *Learning through teaching mathematics: Development of teachers' knowledge and expertise in practice* (pp. 3–21). Dordrecht, Netherlands: Springer.

Lutton, A. (2012). *Advancing the early childhood profession.* Washington, DC: National Association for the Education of Young Children.

Merritt, E., Rimm-Kaufman, S., Berry III, R., Walowiak, T., & McCracken, E. (2010). A reflection framework for teaching math. *Teaching Children Mathematics, 17*(4), 239–246.

National Association for the Education of Young Children & National Association of Early Childhood Specialists in State Departments of Education. (2003). *Early childhood curriculum, assessment, and program evaluation.* Washington, DC: NAEYC.

National Association for the Education of Young Children & National Council of Teachers of Mathematics. (2002). *Early childhood mathematics: Promoting good beginnings.* Washington, DC: Author. Retrieved from http://www.naeyc.org/files/naeyc/file/positions/psmath.pdf

National Board for Professional Teaching Standards. (2010). *The Five Core Propositions.* Retrieved from http://www.nbpts.org/five-core-propositions

National Mathematics Advisory Panel. (2008). *Foundations for success: The final report of the National Mathematics Advisory Panel.* Washington, DC: U.S. Department of Education. Retrieved from http://www2.ed.gov/about/bdscomm/list/mathpanel/report/final-report.pdf

National Research Council. (2009). *Mathematics learning in early childhood: Paths toward excellence and equity.* Committee on Early Childhood Mathematics, C.T. Cross, T.A. Woods, & H. Schweingruber (Eds.). Center for Education, Division of Behavioral and Social Sciences and Education. Washington, DC: National Academies Press.

Palardy, G., & Rumberger, R. (2008). Teacher effectiveness in first grade: The importance of background qualifications, attitudes, and instructional practices or student learning. *Educational Evaluation and Policy Analysis, 30*(2), 111–140.

Philipp, R.A., Ambrose, R., Lamb. L.L.C., Sowder, J.T., Schappelle, B.P., Sowder, J.T., . . . Chauvot, J. (2007). Effects of early field experiences on the mathematical content knowledge and beliefs of prospective elementary school teachers: An experimental study. *Journal for Research in Mathematics Education, 38*(5), 438–476.

Prusaczyk, J., & Baker, P.J. (2011). Improving teacher quality in southern Illinois: Rural access to mathematics professional development (RAMPD). *Planning and Changing, 42*(1–2), 101–119.

Rudd, L.C., Lambert, M.C., Satterwhite, M., & Smith, C.H. (2009). Professional development + coaching = enhanced teaching: Increasing use of math mediated language in preschool classrooms. *Early Childhood Education Journal, 37*(1), 63–69.

Saracho, O., & Spodek, S. (2008). Research perspectives in early childhood mathematics. In O.N. Saracho and B. Spodek (Ed.), *Contemporary perspectives on mathematics in early childhood education* (pp. 309–320). Charlotte, NC: Information Age.

Sarama, J. (2002). Listening to teachers: Planning for professional development. *Teaching Children Mathematics, 9*(1), 36–39.

Sarama, J., & Clements, D.H. (2009). *Early childhood mathematics education research: Learning trajectories for young children.* New York, NY: Routledge.

Scher, L., & O'Reilly, F. (2009). Professional development for K–12 math and science teachers: What do we really know? *Journal of Research on Educational Effectiveness, 2*(3), 209–249.

Seo, K.-H., & Ginsburg, H.P. (2004). What is developmentally appropriate in early childhood mathematics education? Lessons from new research. In D.H. Clements, J. Sarama, & A.-M. DiBiase (Eds.), *Engaging young children in mathematics: Standards for early childhood mathematics education* (pp. 91–104). Hillsdale, NJ: Lawrence Erlbaum Associates.

Tassell, J.L., Marchionda, H., Baker, J., Bemiss, A., Brewer, L., Read, K., Woods, D. (2011). Transformational professional development: Teacher learning through a bifocal lens. *NCSM Journal of Mathematics Education Leadership, 13*(2), 44–51.

Thames, M.H., & Ball, D.L. (2010, November). What math knowledge does teaching require? *Teaching Children Mathematics, 17*(4), 220–229.

Thorton, J., Crim, C., & Hawkins, J. (2009). The impact of an ongoing professional development program on prekindergarten teachers' mathematics practices. *Journal of Early Childhood Teacher Education, 30,* 150–161.

Tirosh, D., Tsamir, P., Levenson, E., & Tabach, M. (2011). From preschool teachers' professional development to children's knowledge: Comparing sets. *Journal of Math Teacher Educators, 14,* 113–131.

Torff, B. (2011). Teacher beliefs shape learning for all students. *Kappan, 93*(3), 21–23.

Wilson, J.H. (2009). *An exploration of early childhood leaders' perceptions regarding their knowledge, skills, and confidence in the areas of mathematical content, child development, pedagogical content, and instructional leadership strategies.* (Unpublished doctoral dissertation). University of Houston, Texas.

5

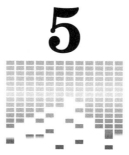

General Features of Effective Professional Development

Implications for Preparing Early Educators to Teach Mathematics

Martha Zaslow

The research on early childhood professional development is burgeoning, and it is beginning to yield conclusions about features of effective professional development. This chapter asks the question: What, from this growing body of work, is important to draw on when seeking to prepare early educators to teach young children mathematics?

This chapter progresses primarily from the broad to the specific: from the research on early childhood professional development in general to implications for teacher preparation specifically in early mathematics (as indicated by the arrow pointing to the right in Figure 5.1). However, there are also important implications for early childhood professional development in general that can be gleaned from seeking to address challenges and opportunities that are specific to preparing early educators to teach mathematics. Accordingly, this chapter concludes with the question of how efforts to prepare teachers in early mathematics can be informative for early childhood professional development more broadly (as indicated by the arrow pointing to the left in Figure 5.1).

The chapter begins by providing background on two reviews of the research on early childhood professional development that serve as the primary resources for this chapter. The chapter discusses six features of effective early childhood professional development that emerge from these reviews. Next, the implications of these features for preparing teachers to instruct young children in math are noted. This discussion includes a cautionary note from the evidence on the status of early childhood professional development, especially inside institutions of higher education. The chapter concludes by asking how research on addressing the challenges specific to professional development in early mathematics can be generalized to strengthen early childhood professional development irrespective of the specific content.

GENERAL ISSUES IN EC PD  ISSUES SPECIFIC TO EC MATH PD

Figure 5.1. Professional development in *early math* draws on and contributes to the broader body of work on early childhood professional development.

RESOURCES DRAWN ON IN THIS CHAPTER

This chapter draws on two reviews of the research on early childhood professional development. The first review, by Zaslow, Tout, Halle, Whittaker, and Lavelle (2010), was conducted as part of a project for the U.S. Department of Education, providing a cross-site evaluation of the Early Childhood Educator Professional Development (ECEPD) programs. The ECEPD programs were funded by the Department of Education, as authorized by Congress, with the aim of strengthening professional development of early childhood educators working in low-income neighborhoods. Multiple cohorts of programs were funded over a period of years, with the Department of Education requiring each program to provide a description of the approach used to strengthen professional development as well as the methodology and results of an evaluation of the program. The cross-site evaluation looked across the evaluations of 18 ECEPD programs, gathering further information for a subset of projects (Tout, Halle, Zaslow, & Starr, 2009).

To provide a foundation for conducting the cross-site evaluation, the project called for conducting a literature review identifying features of effective early childhood professional development that could be used as a framework in reviewing the approaches taken by the ECEPD programs. The literature review, along with some of the findings of the cross-site evaluation, will be drawn upon in identifying key features of effective early childhood professional development.

One of the important features of professional development identified in Zaslow, Tout, Halle, Whittaker, and colleagues (2010) is a direct focus on practice, defined as a focus on observing and providing feedback on interactions with children and the structuring of those interactions through daily routines. Many professional development approaches for early educators historically have aimed at increasing knowledge through college coursework or workshops. An underlying assumption has been that increasing knowledge (such as knowledge of developmental sequences in children's understanding of early literacy or math) will by itself translate into improvements in actual practice.

Yet recent research has raised questions about whether approaches aimed at increasing knowledge ("knowledge-focused professional development" using terminology introduced by Neuman & Cunningham [2009]) on their own result in improvements in overall quality in early childhood classrooms or in gains in achievement for children. Early and colleagues (2006, 2007) found little relationship between early educators' educational attainment and either classroom quality or gain scores in children's achievement in the year prior to kindergarten. In addition, a study by Neuman and Wright (2010) contrasting the effects of college coursework and coaching as approaches to professional development aimed at strengthening early language and literacy development found effects on practice only for coaching.

Particularly in light of such findings, a second key resource is Isner and colleagues' (2011) literature review focusing on coaching. This review looked across the results of 44 studies evaluating professional development approaches involving either on-site or technologically mediated coaching, seeking to determine the consistency of positive results for quality and child outcomes across coaching studies as well as to begin to identify the features of coaching that underlie positive effects. While the general review of the early childhood professional development literature conducted for the ECEPD project and the review of research specifically on coaching form the basis for much of what is discussed in this chapter, we note that because the research on early childhood professional development is progressing very rapidly, we complement summaries from these reviews with findings from individual studies published more recently. In addition, although the following discussion pertains to both preservice and in-service professional development, there are also issues specific to preparation for teaching early math at each of these stages, as discussed in detail by Michael D. Preston regarding preservice professional development (Chapter 6) and by Kimberly Brenneman regarding in-service professional development (Chapter 7).

KEY FEATURES OF EFFECTIVE EARLY CHILDHOOD PROFESSIONAL DEVELOPMENT

The literature review conducted to provide a framework for the cross-site evaluation of ECEPD programs identified six features of early childhood professional development programs that appear to be associated with stronger evidence of effectiveness. These features are briefly described here.

Clearly Articulated and Specific Objectives for Professional Development

It is not surprising that professional development programs or initiatives vary in terms of the breadth of their goals. For example, some professional development programs in early childhood focus on improving the observed quality of early childhood classrooms overall, whereas others aim to improve the stimulation children receive for development only in a specific domain, such as early language and literacy or early mathematics. However, in addition to breadth of goals, research indicates that there is also variation in the degree to which programs of professional development provide clear articulation of specific goals.

There is evidence that professional development approaches with more specific and clearly articulated goals are more effective. Fukkink and Lont (2007) reviewed the evidence on professional development training approaches that focused on strengthening early educators' interaction skills. A meta-analysis of the results of the relevant evaluation studies found significant overall effects of such approaches. Additional analyses contrasted the results of such training approaches when the focus was more specific versus open. The meta-analysis found stronger effects when the goals of the training were more specific.

Greater specification of goals can occur in multiple ways. Teachers and program directors can be introduced to and familiarized with an observational measure of quality, and specific ratings on the measure can be used to set targets for improvements. Research suggests that it may be helpful to go beyond familiarizing early educators with observational measures of quality by providing opportunities for staff members to become proficient in actually rating recorded or live classroom interactions so that they can accurately identify positive practices (Hamre, Pianta, Burchinal, & Downer, 2010). Providing feedback to teachers on their own interactions with children *after* they have honed their skills in identifying specific positive practices may be more efficient and effective.

The issue of specificity of goals may be especially important in preparing teachers in the area of early mathematics. The research on how young children grow in their understanding of math points to not just a specific set of concepts and skills but also their sequencing into learning trajectories (National Research Council [NRC], 2009; Sarama, Clements, Starkey, Klein, & Wakeley, 2008). Thus it may be important in goal setting for professional development in early mathematics to specify both which specific early mathematics concepts and skills are being targeted and how teaching focusing on this set of concepts and skills is part of a sequence, building on earlier learning and laying the foundation for later learning (Clements, Sarama, Spitler, Lange, & Wolfe, 2011).

Direct Focuses on Early Educator Practice

As noted, research is increasingly pointing to the importance of changing early educators' interactions with children by focusing professional development directly on practice. Rather than improvements in knowledge being assumed to lead to changes in practice, the research is increasingly supporting the conclusion that a direct focus on practice is needed in order to improve quality and child outcomes.

This body of research is raising the possibility that introducing general concepts about children's development within a domain such as early literacy or math may work better when the concept is illustrated to the early educator through showing and practicing its application in working with children than through decontextualized discussions. Interesting approaches are emerging in which the focuses on teachers' practice and knowledge are tightly interwoven. Actually seeing the concepts in direct work with children may help teachers gain confidence in the relevance of the concepts for children's development and in their own ability to work with children on these concepts. This may be especially important in the area of early mathematics, where knowledge about specific concepts is central but where teachers may lack the background knowledge and confidence (Ginsburg, Lee, & Boyd, 2008). We now turn our attention to the emergence of practice-focused professional development and then to approaches in which practice- and knowledge-focused professional development are tightly aligned.

Practice-focused professional development is now occurring through onsite individualized coaching and through technologically mediated approaches. In the former, a coach—who is herself an experienced early childhood educator with skills in establishing rapport and helping adults reflect on and improve their educational practices—models positive practices and then observes teachers and provides feedback on their practices. In the latter, a video library provides the models of positive practice, and teachers receive feedback on videos they record in their classrooms and send to their coaches. Practice-focused professional development approaches may also involve a hybrid of on-site and technologically mediated modeling and feedback. For example, Powell, Diamond, and Burchinal (2012), in their work on coaching to strengthen children's oral language skills in early childhood classrooms, have found that initial on-site coaching sessions can be effectively combined with subsequent technologically mediated sessions. They have also carefully piloted the way in which the technologically mediated sessions unfold so that appropriate (and not overwhelming) amounts of information about positive practices and feedback on behavior are provided.

As noted earlier, Isner and colleagues (2011) conducted a review of 44 studies focusing on coaching in early childhood classrooms. Of these studies, 31 examined whether the coaching resulted in improvements in observed quality, with 27 studies (87%) providing some evidence of positive effects. Thus, although there is a fairly consistent pattern of coaching studies showing positive effects, it is important not to assume that all coaching approaches will be effective. A similar pattern was identified in the cross-site evaluation of ECEPD programs (Tout, Halle et al., 2009). All 18 programs in the cross-site evaluation included on-site individualized coaching for the early educators. Of the 18 programs, 10 were found to have evaluations that met criteria for rigor in the way they presented the methodology of the study and in the way the study was conducted. Of these studies, eight showed evidence of positive effects on quality and/or child outcomes. Coaching was present in the programs showing positive effects but did not suffice to assure positive effects.

Such a pattern suggests that it is very important to begin specifying in greater detail which aspects of the various coaching approaches underlie positive effects. Isner and colleagues (2011) concluded that existing studies provide only a limited basis for identifying which features of coaching approaches are most important. Very few studies vary specific dimensions of coaching (such as how much coaching is provided, how coaches are trained or supervised, what specific steps coaches take to model positive practices, or how they provide feedback on early educator practice) and study the outcomes in light of this variation. A new project with funding from the Administration for Children and Families, Office of Planning, Research and Evaluation is laying the groundwork for a study of coaching in Head Start that will seek to systematically vary dimensions of coaching with the goal of learning which dimensions are most important to positive outcomes (Administration for Children and Families, n.d.-b).

Coaching approaches are a frequently used strategy for quality improvement initiatives such as Quality Rating and Improvement Systems (QRIS). QRIS seek to provide parents with readily interpretable summary ratings of quality to inform their choices of early care and education while also providing incentives and supports for quality improvement to early care and education providers (Tout, Zaslow, Halle, & Forry, 2009). A compendium profiling the specific approaches taken in 26 state and local QRIS (Tout et al., 2010) found that each one included on-site assistance to support quality improvements. QRIS generally involve in-service work with teachers, focusing on improving practice with those already in positions working with children. Furthermore, much of the research to date on coaching focuses on quality improvement approaches conducted as part of in-service professional development. Yet there is no reason coaching cannot be directly incorporated into preservice training or coursework, making coursework explicitly practice-focused and helping to combine theory and practice, thus supporting the development of more reflective practitioners. Although we have early promising evidence on coaching as incorporated into college coursework (Hamre et al., 2010), more work is needed evaluating coaching as a component of preservice training to complete licensing requirements or as part of preservice college coursework.

We are beginning to see professional development approaches that not only combine but also carefully align knowledge-focused and practice-focused components. Close linkages between the focus of group training or coursework and of coaching was one of the features of the ECEPD programs showing evidence of effectiveness (Tout, Halle et al., 2009). The recent research of Wasik and Hindman (2011) provides an illustration of effective early childhood professional development with closely aligned group training and individualized coaching. In the professional development approach they report on, group training is provided in modules, each focusing on increasing early educators' understanding of specific aspects of children's language development and including discussion of practices that can support these. Teaching teams develop lesson plans for their classrooms as part of the group training. Coaches then go to individual classrooms, focusing the coaching they provide specifically on the practices related to the topic of the current module. Coaches first model practices related to the topic while teaching teams observe and record instances of positive practice, then teaching teams implement the lesson plan they had developed in the group training while the coach observes and subsequently provides feedback. As noted, approaches that tightly align practice-focused and knowledge-focused components may be helpful to consider in teacher preparation in early mathematics.

Collective Participation in Professional Development of Teachers and Other Staff from the Same Classroom, Program, or School

There is substantial variation across professional development approaches in whether only the lead teacher in a classroom participates, whether the entire classroom teaching team (including an assistant teacher) participates, or whether all program staff participate. Most of the coaching studies included in the review by Isner and colleagues (2011) described coaching approaches that focused only on the lead teacher or a family child care provider. However, a number of recent studies describe coaching approaches that encompass a lead teacher and the assistant teacher or aide. For example, the Chicago School Readiness Project (Raver et al., 2011), which aimed to enhance children's social and emotional development with a particular focus on self-regulation, included lead and assistant teachers together in both training and coaching. The Foundations of Learning program (Morris, Raver, Millenky, Jones, & Lloyd, 2010), which focused on improving classroom management of children's behavior, also targeted lead and assistant teachers. There is less discussion of teams participating in group training or coursework, but some examples have been described in the literature (e.g., Ginsburg et al., 2006).

Inclusion of all staff members in a program—not only teaching teams but also directors, specialists, and support staff—tends to occur when the goals of quality improvement are at the program level as well as the classroom level. Such approaches may be especially important for sustaining the positive effects of professional development. Thus, for example, a study by Shlay and colleagues (2001) described a coaching approach aimed at supporting programs' work toward accreditation that included directors along with classroom staff. With QRIS also involving ratings at the program level, we are also seeing descriptions of coaching as part of QRIS quality improvement efforts that include directors as well as classroom staff. However, Isner and colleagues' (2011) case studies of four local QRIS found variation in the extent to which the program director was included in the coaching. *Not* including the director in the coaching appeared to carry with it the possibility of the director not fully understanding the coaching program and not conveying support for it in guidance to teachers, multiple coaching approaches being used within a program without coordination, and directors feeling undermined by coaches coming to work with staff members without their involvement. QRIS seek to improve quality across early childhood programs occurring in the full range of settings, including early childhood classrooms within public schools. Thus this research suggests the need to consider the issue of how best to inform and include school principals as well as directors of nonschool-based early childhood programs in coaching approaches.

We do not, as yet, have studies that systematically vary whether professional development focuses only on the lead teacher or encompasses other staff members as well. However, the literature review conducted for the ECEPD project found some indication of greater effectiveness when professional development approaches encompassed more staff members. The involvement of teaching teams, of staff at different levels of a program (e.g., directors as well as teachers), and also of staff working with children of different ages in a program can foster programwide consistency in practices. In contrast, a single teacher who has received professional development may be introducing practices into her classroom that other teachers or staff members do not understand or feel supportive of, and may even contradict. Including multiple staff members in the same professional development can also provide a starting point for reflective practice by groups of staff members, something that may be especially important to sustaining changes in practice, especially in areas that many teachers see as new and challenging, such as mathematics teaching.

Much more work is needed to provide guidance on how best to engage multiple staff members in coaching. For example, work by Allard Agnamba (2012) raises the question of whether those who are skilled in building relationships with individual teachers or teaching teams in a coaching relationship are necessarily the best staff members to work directly with principals or program directors. Different skills appear to be needed in these relationships. Furthermore, when models of positive practices are provided for technologically mediated coaching, the research to date generally describes videos of lead teachers alone. An important next step in the work on early childhood professional development is to provide more specific guidance to programs on how best to incorporate staff members other than lead teachers. Video examples of teaching teams working together would also be extremely valuable. These could be used in both preservice and in-service teacher preparation (see Chapters 6 and 7).

Alignment of Intensity and Duration of Professional Development with the Content Covered

The total "dosage" of a professional development approach may include the duration of each class or coaching session, the frequency and total number of such sessions, and the duration of the entire series of sessions. Different professional development approaches vary substantially in overall dosage, as well as how the overall dosage is arrived at through the frequency and number of sessions and their duration. For example, in their review of coaching studies, Isner and colleagues (2011) found the frequency of reported coaching to range from once a month to multiple times per week, with some coaching sessions lasting fewer than 90 minutes and others longer, and with coaching continuing from fewer than 2 months to longer than a year. Some recent studies of coaching underscore the need for a longer duration (2 years) in order for changes to be not only attempted but also solidified and sustained (Wasik & Hindman, 2011). Parallel variation exists in the frequency of classes involved in coursework or training, the length of each individual class, and the number of weeks classes continue. Approaches that include both knowledge-focused and practice-focused professional development will need to take into account the cumulative dosage of these components.

A key conclusion from the literature review conducted for the ECEPD project is that dosage needs to match the breadth of content being conveyed and the depth in which it is conveyed. Positive effects have been found for professional development that involved a limited dose when the specified goal of professional development was mastery by early educators of a single activity or strategy to use with children (see, for example, Whitehurst et al., 1994). However, a small dose of professional development is not appropriate if the goal is to convey theory and practice across multiple aspects of development in a domain (e.g., oral language, phonological awareness, alphabetic principle, and awareness of print in early language and literacy development) or to strengthen development in multiple domains. Given the current level of knowledge, skill, and beliefs in early childhood mathematics teaching (Hyson & Woods, Chapter 2), it would seem that larger doses of professional development are essential.

Another key issue is the distinction between dosage as intended and dosage as actually received. Evidence that dosage as actually received is important comes from a study by Mashburn, Downer, and Hamre (2010) regarding teacher use of technologically mediated coaching in a professional development approach called MyTeachingPartner (for further discussion of this professional development approach, see Brenneman, Chapter 7; Vick Whittaker & Hamre, Chapter 8). This study found that children's gains in vocabulary during a year of preschool were greater when their teachers spent more time engaged in the online coaching.

Further work is needed examining effects on teachers and children when the dosage of coaching is systematically varied. To date, the research suggests that calibration is needed between the scope of the goals of professional development and dosage and that it is important to assure that the intended dosage is actually received. Another important issue related to dosage is cost effectiveness. For example, there may be reluctance to put into place a 2-year course of professional development. Yet, if the evidence shows that a larger dose is needed to bring about sustained change, this larger and more expensive initial dose may actually be cost effective. In the area of early mathematics, where, as we have noted, the focus is not on children's mastery of isolated concepts but on concepts that build sequentially over time, it again seems particularly critical to look at dosage relative to sustained effects over time.

Preparation for Conducting and Using the Information from Child Assessments

The reviews of the evidence for the ECEPD project suggests that effective professional development programs tend to incorporate training for early educators not only in conducting child assessments but in using the information from assessments to monitor children's progress and inform instruction. It is quite different to simply be able to administer or complete the observations needed for child assessments than it is to use the information obtained. Professional development on both is needed.

Such professional development is highly dependent on the availability of appropriate assessment tools. In this instance, the tools that are needed help pinpoint the progress a child has made in the development of specific concepts and skills as well as support and inform teachers' targeting of instruction. Such tools differ from tools needed for evaluation or for providing a descriptive picture of children's development across a geographical region. For example, in assessment aimed at guiding instruction, it is appropriate to probe or repeat an item so that a teacher is sure he or she has not missed capturing a child's level of understanding or skill. However, in assessment for evaluation or to describe the educational status of children in a region, the assessment must be conducted in a strictly standardized manner (Snow & Van Hemel, 2008).

We actually have very little information on how programs help to prepare their staff to conduct child assessments or how programs use the information from assessments to monitor children's progress and inform instruction. The Office for Planning, Research and Evaluation in the Administration for Children and Families, part of the U.S. Department of Health and Human Services, has recently launched a project on early educators' use of progress monitoring approaches to individualize teaching (Administration for Children and Families, n.d.-a).

The use of assessments to inform instruction may be particularly important in areas such as early mathematics, where there are sequences of learning such that one skill is foundational for mastering subsequent skills (see Ginsburg, 2009; Ginsburg, Chapter 3), and where teachers often know little about children's mathematical development (see Hyson & Woods, Chapter 2). Frustration for both child and teacher, and lack of progress in children, can result from instruction where foundational understanding is lacking. Using assessments to diagnose a child's level of understanding can eliminate such frustration and strengthen learning.

Just as there can be lack of tight alignment between knowledge-focused and practice-focused professional development, there can also be a lack of alignment in professional development on assessments and instructional practices. Integrating these elements of professional development is critical so that teachers understand how to both assess whether a child has mastered a concept or skill and then build to the next step once the child is ready.

Consideration of Organizational Context as Well as Standards for Practice

A growing body of evidence indicates that the strength of any approach to improving quality in early childhood programs will reflect both the effectiveness of the initial approach as implemented in tightly controlled circumstances (e.g., in a demonstration project) and the way in which it is implemented on a larger scale (Halle, Metz, & Martinez-Beck, 2013). Implementation, in turn, reflects both organization-level factors and broader systems-level factors (Brenneman, Chapter 7). Failure to take organizational- and systems-level factors into account can result in a program failing to show the effects it had in a demonstration project when it is brought to scale (see the earlier discussion regarding the inclusion of directors and principals as well as other staff members in professional development).

For early childhood professional development, key organizational factors related to implementation will include the extent to which the professional development approach being introduced is in harmony with a curriculum that is already in place and the approaches taken by specialists such as curriculum specialists or family support workers. A new approach to professional development may also flounder if it does not take into account key characteristics of the population in a classroom or program, such as the concentration of dual-language learners.

In Chapter 1, Kagan and Gomez point to important systems-level factors that can facilitate or hinder the implementation of a program of professional development. For example, if the focus of the professional development does not align well with a state's early learning guidelines for early mathematics or build appropriately toward Common Core standards for K–12, it will likely not be given the attention and support that is needed for full implementation nor will it result in positive outcomes on assessments that align with these guidelines and standards. Similarly, if the state has developed or is in the process of working toward a kindergarten entry assessment as part of its work for Race to the Top/Early Learning Challenge funding, the way in which the assessment seeks to capture children's skills in early mathematics will be very important to the implementation of any particular professional development approach.

IMPLICATIONS FOR PROFESSIONAL DEVELOPMENT IN TEACHING EARLY MATHEMATICS

In what ways can teacher preparation in the area of early childhood mathematics build on this set of features? Preparing early childhood teachers for instruction in mathematics fits readily with some of these features, whereas for others there appear to be challenges. This section first turns to the general features of early childhood professional development that appear to align well with needs in preparing teachers in this specific content area and then turns to areas where greater attention is likely to be needed.

Features that Fit Well with Preparation of Early Educators to Teach Mathematics

Specificity of Goals

Early mathematics is unusual among the content areas that contribute to young children's school readiness because of the clear articulation of the central concepts and skills young children already possess and those they need to master to provide a strong foundation for later learning (NRC, 2009). This framework provides a critical starting point for the clear articulation of specific goals both for children's learning and for preparing teachers for instructing children so that they will progress toward these goals.

To briefly recapitulate what is discussed elsewhere in this volume in greater detail, there was a longstanding hesitance to provide intentional instruction in mathematics in the early years because such instruction was seen as developmentally inappropriate for young children and requiring didactic approaches. However, a substantial body of research (as summarized by the National Academy of Science Committee on Early Childhood Mathematics; NRC, 2009) indicates that understanding of math concepts begins at a very early age. The research also clearly indicates that learning of early mathematics concepts and skills can be fostered in interesting and engaging ways by instructional approaches and organized curricula appropriate for young children. Furthermore, understanding of concepts in early mathematics is among the strongest predictors of later achievement (Duncan et al., 2007). Therefore it is particularly of concern that discrepancies in early mathematics achievement by socioeconomic status emerge before kindergarten (Klibanoff, Levine, Huttenlocher, Vasilyeva, & Hedges, 2006). Fostering early math skills, especially among children at risk in terms of school readiness, needs to be a high priority.

The framework articulated by the Committee on Early Childhood Mathematics has fostered the identification of clear goals for the professional development of early educators, as is evident in Chapter 3 by Ginsburg—who served as one of the committee members and worked closely with Woods, the Study Director, as well as Hyson, who prepared a commissioned paper for the committee (Hyson, 2008). Other domains of children's school readiness do not have such clearly articulated frameworks. In the area of social and emotional development, for example, there are multiple and somewhat differing conceptualizations and articulations of key constructs (Hyson et al., 2011). The existence of an overarching framework for early mathematics concepts and skills and an understanding of how early math skills build progressively is a key resource to draw upon in addressing the feature of effective professional development of specificity in articulating goals. The availability of a framework does not, however, eliminate the need for substantial work in developing, testing, and refining comprehensive programs of professional development based on the framework.

Calibration of Dosage to Goals

This chapter has noted that in effective early childhood professional development, dosage is calibrated to the breadth and depth of content on which early educators are being prepared to provide instruction. In the review of the literature conducted for the ECEPD project, it was clear that some early mathematics interventions focus only on limited aspects of the overarching conceptualization of early math skills (e.g., Sophian, 2004, who focused specifically on the use of differing unit sizes in measurement and the implications of this for numerical outcomes), whereas others address a range of early childhood mathematics skills (e.g., Starkey, Klein, & Wakeley, 2004, who focused on enumeration and number sense, arithmetic reasoning, spatial sense, geometric reasoning, pattern sense and unit construction, nonstandard measurement, and logical relations). Thus calibration of dosage of professional development can occur in relation to the extent of the overarching framework that professional development is seeking to address. An important underlying question is whether seeking to teach children only selected aspects from the framework will suffice to lay a strong foundation for later learning in mathematics.

Direct Focus on Practice

As described by Hyson and Woods in Chapter 2, research points to hesitance or even resistance by some early educators to teach mathematics, based at least in part on lack of confidence in their own understanding of key concepts. Interestingly, the emerging evidence that effective early childhood professional development focuses on practice, and not simply knowledge, may help to allay such concerns. Just as some of the terminology for the key concepts in children's language development (e.g., phonological awareness) can sound jargony, terminology such as number sense in early childhood mathematics can also be off-putting.

Rather than introducing the important concepts for the development of early math skills in knowledge-focused approaches, the research on effective early childhood professional development suggests that it may be more effective in changing practice to emphasize and even begin with practice-focused approaches. Helping teachers to see the concepts in action may provide a starting point for understanding and discussing the concepts in group training or coursework. Indeed, although it was long assumed that knowledge of a concept would lead to successful practice, an alternative conceptualization is that enacting and reflecting on positive practices provides an effective basis for deriving and understanding the underlying concepts (Zaslow, Tout, Halle, & Starr, 2010). Professional development in early mathematics may provide a crucial context in which to explore how best to interweave the introduction of positive practices and mastery of concepts related to children's development.

Features with Which There Are Likely to Be More Significant Challenges

Collective Participation in Professional Development

This chapter notes the suggestive evidence that collective participation in professional development by teaching teams or all staff members in a program may result in greater change in behavior, and perhaps more sustained change, than professional development addressed to only a lead teacher. Yet the chapter also notes that little guidance is available for how to effectively engage directors or principals in professional development, including whether the same professional development provider should work with a director as well as the teaching staff. Similarly, although a number of professional development approaches have been addressed to teaching teams, we lack written illustrations or a library of videos showing how the members of the team can be prepared to work in a coordinated way in providing instruction.

These gaps may pose a particularly significant challenge for professional development in early mathematics. The literature on implementation of early math curricula includes descriptions of directors who do not see the need to instruct young children in math, and who may not support, or who may even undermine, staff undergoing professional development to implement an early math curriculum (see, for example, Ginsburg et al., 2006). We urgently need to address the lack of guidance on how to include directors of early childhood programs, principals, and others responsible for program administration and oversight (such as master teachers and curriculum specialists) in professional development so that they are informed about and supportive of the preparation of their staff to teach mathematics.

Conducting and Using Information from Assessments

The lack of clear illustrations of teaching teams working together on the implementation of a curriculum may be especially problematic in a content area where individualization is so clearly called for. If mastery of one skill provides the needed foundation for the next, instruction in early math needs to take into account the progress individual children have made in order to set goals for next steps. Yet assessing individual children's progress and setting individualized learning goals is extremely difficult if the lead teacher alone is responsible for planning and conducting learning activities and assessments. Key decisions will need to be made about preparation for observation and assessment of children. Should both the lead teacher and assistant teacher monitor children's progress and therefore be prepared on recording observations or conducting assessments? If so, will staffing patterns ensure that assistant teachers are assigned to the same classroom over time so that they are available to monitor individual children's progress? How will such work in teams take into account and overcome the challenges of turnover among both lead and assistant teachers (see Kagan & Gomez, Chapter 1)?

Professional Development in the Context of Systems

A further challenge in professional development for early mathematics is that some key systems-level factors are currently in flux. A particularly important issue is that intensive work is currently in progress in multiple states to develop kindergarten entry assessments (see Kagan & Gomez, Chapter 1). It is impossible to align instruction in early mathematics with the measurement of early math skills when the measurement is still under development.

A CAUTIONARY NOTE: GAPS IN PRESERVICE AND IN-SERVICE PROGRAMS

Before concluding this chapter, it seems imperative to interject a cautionary note. A recent review of the evidence on professional development for early educators provided through programs of higher education by Hyson, Horm, and Winton (2012) raises questions about the extent to which the key features of effective professional development for early educators noted here are seen in higher education programs.

Hyson and colleagues (2012) summarize evidence that math receives less coverage than early literacy in associate's and bachelor's degree programs for early educators. For example, whereas 65% of associate's degree programs were found to have coursework in early literacy, 49% had coverage for early math. The parallel figures for bachelor's level programs are 77% and 59%. In surveys, faculty members themselves identify gaps in their knowledge base in early mathematics. Furthermore, vague responses by faculty members to survey questions about what they teach their students regarding math curricula for young children, and what math-related competencies they expect their students to show, raise concern about the specificity of goals in professional development focusing on early mathematics provided in higher education.

The review by Hyson and colleagues (2012) also raises concerns about the extent to which higher education programs include an explicit focus on practice and also tight linkages between knowledge-focused and practice-focused professional development. Hyson and colleagues (2012) conclude that teacher preparation programs for early childhood educators in institutions of higher education provide limited attention to practice, focusing more heavily on the acquisition of knowledge (setting aside the key issue of the extent to which this knowledge is current according to recent research). Summarizing a study by Johnson, Fiene, McKinnon, and Babu (2010), Hyson and colleagues (2012) note limited intentional linking of coursework focusing on specific content areas (such as early science or mathematics) and field placement experiences involving observation and feedback on practice in the particular content area. Hyson and colleagues (2012) underscore the need for research on faculty skills in fostering student ability to implement specific practices.

Thus there is cause for concern about the extent to which early childhood education degree programs are incorporating the general approaches supported by the research on effective professional development, especially those related to specificity and focus on practice. We do not have a parallel review of the evidence that would permit an assessment of the extent to which training workshops that do not contribute to a higher education degree, or coaching conducted to address licensing requirements or for QRIS, encompass the features noted. However, just as for professional development provided through higher education, we have some troubling indications that professional development carried out outside of institutions of higher education does not always include all the key features of effective professional development discussed in this chapter (Zaslow, Tout, Halle, Whittaker et al., 2010).

One issue appears to be the degree to which on-site individualized professional development through in-service coaching is aligned with current systems for recognizing levels of professional development. A teacher's participation in coaching to improve quality for a QRIS rating may not be accorded official recognition in terms of how the same QRIS marks progress on the teacher's professional development. In short, participation in coaching, even if it yields improvements in quality, may not contribute to in-service training requirements. As one example, in Washington State's Seeds for Success evaluation (Boller et al., 2010), even though coaching resulted in observed improvements in classroom quality, improvements on QRIS ratings of professional development were limited because the coaching did not contribute to improved ratings—that is, the system provided no way to recognize the coaching in existing markers of progress on professional development.

In sum, in order to make progress in early educators' preparation to teach mathematics, systems of professional development, both preservice and in-service, will need to be structured in ways that support such efforts.

CONCLUSION

This chapter has progressed from broad to specific: from a discussion of general features of effective early childhood professional development to the identification of implications for preparing early childhood teachers for instruction in the specific area of mathematics. However, the relationship is reciprocal. Growth in our understanding of how to prepare teachers for the instruction of young children in mathematics stands to yield insights relevant to our general understanding of effective professional development. The following are examples:

- Testing whether starting with a focus on practice and using that starting point to work toward knowledge of actionable concepts in early mathematics helps to allay teacher concerns about understanding the underlying concepts in math. This would suggest important possibilities for mastering concepts in other domains of development.

- Developing case studies of effective preparation of directors, principals, and other education leaders so that they understand and support professional development in early mathematics could pave the way for approaches to work with education leaders in preparation for professional development in other key domains or in integrative cross-domain curricula (see the discussion of integrated curricula in Chapter 7).

- Building a video library of lead and assistant teachers working together to implement an early math curriculum could yield insights about how to provide professional development for teaching teams irrespective of the content area.

- Similarly, a video library could be developed using interviews and assessments with children to help identify their understanding of key mathematical concepts. Such a video library would be useful for the design of professional development on using assessment to guide individualized instruction, a key issue across developmental domains.

- Grappling with how to align professional development in early mathematics with a kindergarten entry assessment, which is still in development, could yield improved strategies for collaborative participation in the development, piloting, and refinement of the assessment. Such work could inform future steps to tighten alignment of kindergarten entry assessments and early childhood professional development across domains.

- Finally, there are important opportunities for intentional collaboration in professional development for instruction in multiple domains of early childhood development. For example, vocabulary development can be intentionally fostered in instruction in early mathematics. Similarly, expressive language can be fostered in teacher interactions with children as the children work on solving problems in math activities. Integrative goals for early childhood professional development build on an understanding of the interrelated nature of children's development across domains (Sarama, Lange, Clements, & Wolfe, 2011). They also hold the potential of diminishing overload when teachers receive professional development in multiple content areas (Bouffard & Jones, 2011; Brenneman, Chapter 7).

Strengthening professional development in early mathematics can both build on and also contribute in important ways to our understanding of the general features of effective early childhood professional development.

FOR REFLECTION AND ACTION

1. Zaslow summarizes two recent reviews of research on early childhood professional development. These reviews are well worth reading as background, especially if you are a researcher or leader of new professional development efforts in early childhood mathematics education. If possible, discuss their implications with colleagues.
2. A recurring theme in research on effective professional development is that there needs to be a strong focus on practice. Thinking about your experiences and observations, is there enough of a practice focus? If not, what seem to be the barriers?
3. As Zaslow notes, coaching has become a frequently implemented way of improving early childhood teachers' effectiveness. Again, what has been your experience with coaching, and how successful has coaching been (in math or in other domains)?
4. For your experience as a provider or leader in professional development for early educators, what are the main lessons you draw from Zaslow's discussion? What changes may you wish to make, based on what you have learned from this chapter?
5. As a sort of needs assessment, it might be helpful to design a checklist or rating scale to assess your organization's current professional development efforts in mathematics, using the six key features identified in this chapter.

REFERENCES

Administration for Children and Families, Office of Planning, Research, and Evaluation. (n.d.-a). *Early childhood teachers' use of progress monitoring to individualize teaching practices, 2012–2013.* Retrieved from http://www.acf.hhs.gov/programs/opre/research/project/early-childhood-teachers-use-of-progress-monitoring-to-individualize

Administration for Children and Families, Office of Planning, Research, and Evaluation. (n.d.-b). *Head Start coaching study: Design phase, 2012–2013.* Retrieved from http://www.acf.hhs.gov/programs/opre/research/project/head-start-coaching-study-design-phase

Allard Agnamba, L.T. (2012). *Preparation and ongoing support for early childhood instructional coaches: A case study exploration of an instructional coaching program.* (Unpublished doctoral dissertation). University of Pennsylvania, Philadelphia, PA.

Boller, K., Del Grosso, P., Blair, R., Jolly, U., Fortson, K., Paulsell, D., . . . Kovac, M. (2010). *The Seeds to Success modified field test: Findings from the impact and implementation studies.* Princeton, NJ: Mathematica Policy Research.

Bouffard, S.M., & Jones, S.M. (2011). The whole child, the whole setting: Toward integrated measures of quality. In M. Zaslow, I. Martinez-Beck, K. Tout, & T. Halle (Eds.), *Quality measurement in early childhood settings* (pp. 281–295). Baltimore, MD: Paul H. Brookes Publishing Co.

Clements, D.H., Sarama, J., Spitler, M.E., Lange, A.A., & Wolfe, C.B. (2011). Mathematics learned by young children in an intervention based on learning trajectories: A large-scale cluster randomized trial. *Journal for Research in Mathematics Education, 42*(2), 127–166.

Duncan, G.J., Dowsett, C.J., Claessens, A., Magnuson, K., Huston, A.C., Klebanov, P., . . . Japel, C. (2007). School readiness and later achievement. *Developmental Psychology, 43*(6), 1428–1446.

Early, D., Bryant, D., Pianta, R., Clifford, R.M., Burchinal, M., Ritchie, S., . . . Barbarin, O. (2006). Are teachers' education, major, and credentials related to classroom quality and children's academic gains in pre-kindergarten? *Early Childhood Research Quarterly, 21*(2), 174–195.

Early, D.M., Maxwell, K.L., Burchinal, M., Alva, S., Bender, R.H., Bryant, D., . . . Zill, N. (2007). Teachers' education, classroom quality, and young children's academic skills: Results from seven studies of preschool programs. *Child Development, 78*(2), 558–580.

Fukkink, R., & Lont, A. (2007). Does training matter? A meta-analysis and review of caregiver training studies. *Early Childhood Research Quarterly, 22*(3), 294–311.

Ginsburg, H.P. (2009). The challenge of formative assessment in mathematics education: Children's minds, teachers' minds. *Human Development, 52*(2), 109–128.

Ginsburg, H.P., Kaplan, R.G., Cannon, J., Cordero, M.I., Eisenband, J.G., Galanter, M., & Morgenlander, M. (2006). Preparing Early Childhood Educators to Teach Mathematics. In M. Zaslow & I. Martinez-Beck (Eds.), *Critical issues in early childhood professional development* (pp. 171–202). Baltimore, MD: Paul H. Brookes Publishing Co.

Ginsburg, H.P., Lee, J.S., & Boyd, J.S. (2008). Mathematics education for young children: What it is and how to promote it. *Society for Research in Child Development Social Policy Report, 22*(1), 3–22.

Halle, T., Metz, A., & Martinez-Beck, I. (Eds.). (2013). *Applying implementation science in early childhood programs and systems.* Baltimore, MD: Paul H. Brookes Publishing Co.

Hamre, B.K., Pianta, R.C., Burchinal, M., & Downer, J.T. (2010, March). *A course on supporting early language and literacy development through teacher-child interaction: Effects on teacher beliefs, knowledge and practice.* Presentation at the Meetings of the Society for Research in Educational Effectiveness, Washington, DC.

Hyson, M. (2008). *Preparing teachers to promote young children's mathematical competence.* Paper commissioned by the Committee on Early Childhood Mathematics. Washington, DC: National Research Council.

Hyson, M., Horm, D.M., & Winton, P.J. (2012). Higher education for early childhood educators and outcomes for young children: Pathways toward greater effectiveness. In R. Pianta, L. Justice, S. Barnett, & S. Sheridan (Eds.), *Handbook of early education* (pp. 553–583). New York, NY: Guilford Press.

Hyson, M., Whittaker, J.V., Zaslow, M., Leong, D., Bodrova, E., Hamre, B., & Smith, S. (2011). Measuring the quality of environmental supports for young children's social and emotional competence. In M. Zaslow, I. Martinez-Beck, K. Tout, & T. Halle (Eds.), *Quality measurement in early childhood settings* (pp. 105–134). Baltimore, MD: Paul H. Brookes Publishing Co.

Isner, T., Tout, K., Zaslow, M., Soli, M., Quinn, K., Rothenberg, L., & Burkhauser, M. (2011). *Coaching in early care and education settings.* Report prepared for Children's Services Council of Palm Beach County. Washington, DC: Child Trends.

Johnson, J., Fiene, R., McKinnon, K., & Babu, S. (2010). *A study of ECE pre-service teacher education at major universities in 38 preK states.* Final report to the Foundation for Child Development. State College: Pennsylvania State University.

Klibanoff, R.S., Levine, S.C., Huttenlocher, J., Vasilyeva, M., & Hedges, L.V. (2006). Preschool children's mathematical knowledge: The effect of teacher "math talk." *Developmental Psychology, 42*(1), 59–69.

Mashburn, A.J., Downer, J.T., & Hamre, B.K. (2010). Consultation for teachers and children's language and literacy development during pre-kindergarten. *Developmental Science, 14*(4), 179–196.

Morris, P., Raver, C.C., Millenky, M., Jones, S., & Lloyd, C.M. (2010). *Making preschool more productive: How classroom management training can help teachers.* New York, NY: Manpower Demonstration Research Corporation.

National Research Council. (2009). *Mathematics learning in early childhood: Paths toward excellence and equity.* Committee on Early Childhood Mathematics, C.T. Cross, T.A. Woods, & H. Schweingruber (Eds.). Center for Education, Division of Behavioral and Social Sciences and Education. Washington, DC: National Academies Press.

Neuman, S.B., & Cunningham, L. (2009). The impact of professional development and coaching on early language and literacy instructional practices. *American Educational Research Journal, 46*(2), 532–566.

Neuman, S.B., & Wright, T. (2010). Promoting language and literacy development for early childhood educators: A mixed-methods study of coursework and coaching. *Elementary School Journal, 111*(1), 63–86.

Powell, D.R., Diamond, K.E., & Burchinal, M.R. (2012). Using coaching-based professional development to improve Head Start teachers' support of children's oral language skills. In C. Howes, B.K. Hamre, & R.C. Pianta (Eds.), *Effective early childhood professional development: Improving teacher practice and child outcomes* (pp. 13–29). Baltimore, MD: Paul H. Brookes Publishing Co.

Raver, C.C., Jones, S.M., Li Grining, C., Zhai, F., Bub, K., & Pressler, E. (2011). CSRP's impact on low income preschoolers' preacademic skills: Self regulation as a mediating mechanism. *Child Development, 82*(1), 362–378.

Sarama, J., Clements, D.H., Starkey, P., Klein, A., & Wakeley, A. (2008). Scaling up the implementation of a pre-kindergarten mathematics curriculum: Teaching for understanding with trajectories and technologies. *Journal of Research on Educational Effectiveness, 1*(2), 89–119.

Sarama, J., Lange, A., Clements, D.H., & Wolfe, C.B. (2012). The impacts of an early mathematics curriculum on oral literacy and language. *Early Childhood Research Quarterly, 27*(3), 489–502.

Shlay, A.B., Jaeger, E., Murphy, P., Shaw, K., Gottesman, L., & Weinraub, M. (2001). *Making a case for child care: An evaluation of a Pennsylvania-based intervention called Child Care Matters.* Philadelphia, PA: Temple University, Center for Public Policy.

Snow, C., & Van Hemel, S. (Eds.), Committee on Developmental Outcomes and Assessments for Young Children, National Research Council of the National Academies of Science. (2008). *Early childhood assessment: Why, what and how.* Washington, DC: National Academies Press.

Sophian, C. (2004). Mathematics for the future: Developing a Head Start curriculum to support mathematics learning. *Early Childhood Research Quarterly, 19*(1), 59–81.

Starkey, P., Klein, A., & Wakeley, A. (2004). Enhancing young children's mathematical knowledge through a pre-kindergarten mathematics intervention. *Early Childhood Research Quarterly, 19*(1), 99–120.

Tout, K., Halle, T., Zaslow, M., & Starr, R. (2009). *Evaluation of the Early Childhood Educator Professional Development Program: Final report.* Prepared for the Policy and Program Studies Service, Office of Planning, Evaluation and Policy Development, U.S. Department of Education, Washington, DC.

Tout, K., Starr, R., Moodie, S., Soli, M., Kirby, G., & Boller, K. (2010). *Compendium of quality rating systems and evaluations.* Washington, DC: U.S. Department of Health and Human Services, Administration for Children and Families, Office of Planning, Research and Evaluation.

Tout, K., Zaslow, M., Halle, T., & Forry, N. (2009). *Issues for the next decade of quality rating and improvement systems.* (OPRE Issue Brief No. 3). Washington, DC: Office of Planning, Research and Evaluation, Administration for Children and Families, U.S. Department of Health and Human Services, and Child Trends. Retrieved from http://www.acf.hhs.gov/programs/opre/resource/issues-for-the-next-decade-of-quality-rating-and-improvement-systems

Wasik, B.A., & Hindman, A.H. (2011). Improving vocabulary and pre-literacy skills of at-risk preschoolers through teacher professional development. *Journal of Educational Psychology, 103*(2), 455–469.

Whitehurst, G.J., Arnold, D.S., Epstein, J.N., Angell, A.L., Smith, M., & Fischel, J.E. (1994). A picture book reading intervention in day care and home for children from low-income families. *Developmental Psychology, 30*(5), 679–689.

Zaslow, M., Tout, K., Halle, T., & Starr, R. (2010). Professional development for early educators: Reviewing and revising conceptualizations. In S. Neuman & D. Dickinson (Eds.), *Handbook of early literacy research* (Vol. 3, pp. 425–434). New York, NY: Guilford Press.

Zaslow, M., Tout, K., Halle, T., Whittaker, J.V., & Lavelle, B. (2010). *Towards the identification of features of effective professional development for early childhood educators.* Prepared for Policy and Program Studies Service, Office of Planning, Evaluation and Policy Development, U.S. Department of Education. Retrieved from http://www.ed.gov/about/offices/list/opepd/ppss/reports.html

6

Promising Approaches to Early Childhood Mathematics Education Professional Development in Preservice Settings

Technology as a Driver

Michael D. Preston

This chapter focuses on promising new approaches to teacher preparation for early childhood mathematics education (ECME). The ideas are simple and resonate with other chapters: Teacher candidates in higher education programs need opportunities to learn mathematics, to discover what children know and can do, and to develop their teaching skills by observing expert practice and honing their own practice with adequate feedback and support.

As emphasized in Chapters 1 and 2 and in recent literature reviews (Hyson, Horm, & Winton, 2012), the preparation of future early childhood educators in colleges and universities has shared many of teacher education's more general problems but has also faced some unique challenges. Although many students in early childhood programs begin higher education with years of experience in child care, Head Start, and other settings, many of them have not seen or implemented effective, intentional, and developmentally appropriate practices in their own classrooms (and, judging by national studies, their classrooms in general may not reflect high-quality practices, for example, Mashburn et al., 2008; NICHD Early Child Care Research Network, 2000). Whether or not they have prior teaching experience, once these students enter a college or university program (in either community colleges or four-year institutions), most early childhood teacher candidates take too few math and math education courses, their content knowledge lacks depth and relevance, and they have insufficient math-related field experiences (Hyson & Woods, Chapter 2). Finally, and again as noted in other chapters, early childhood faculty lack—and often acknowledge they lack—the background to prepare their students to provide quality mathematics instruction.

At least in part, these challenges can be addressed by new approaches that leverage technologies to change the way early childhood teacher candidates access and interact with content, including mathematics, pedagogy, and developmental psychology, and to support their development as professionals and adult learners while they build relationships with peers and mentors. Certainly, technology is not the answer to all the many challenges of teacher education, but it can serve as a powerful tool in improving prospective teachers' ability to teach mathematics to young children.

This chapter reviews the current state of preservice programs, highlights several promising new developments in mathematics-focused technologies, and offers a framework and strategies for using these technologies, especially digital video, to support ECME teacher preparation.

THE CURRENT STATE OF PRESERVICE PROGRAMS IN EARLY CHILDHOOD MATHEMATICS EDUCATION

Teacher preparation is at a crossroads, embedded within a larger crossroad looming over institutions of higher education across the country, where costs are increasing and students perceive a declining relevance and return on investment. Regarding teacher education programs, Darling-Hammond notes that

> teachers go into debt to enter a career that pays noticeably less than their alternatives—especially if they work in high-poverty schools—and reach the profession through a smorgasbord of training options, from excellent to awful, often followed by little mentoring or help. (2011)

These factors do constitute a crisis, but they also present an opportunity to review the current state of preservice teacher education, rethink the approach, and design more effective learning experiences for our future teaching work force.

The Challenges of Adequacy and Authenticity in Teacher Preparation

Two primary critiques of teacher preparation in general, including early childhood teacher preparation, are its lack of *adequacy* and *authenticity*. To make teacher preparation more adequate, institutions need to improve teacher candidates' depth of content knowledge, pedagogical content knowledge, and child development. To achieve greater authenticity, they need to provide candidates with more frequent opportunities to participate in real classrooms, first as observers and subsequently as student teachers with increasing amounts of independence and responsibility. Any new approaches—in all areas but certainly including those targeting early childhood mathematics education—should attempt to address both problems.

First, early childhood programs must address the *adequacy challenge* by ensuring their courses and field experiences thoroughly prepare teacher candidates in specific content areas and develop their understanding of the children they will teach and the appropriate methods for teaching them. With respect to mathematics, most early childhood programs tend to place much greater emphasis on literacy than math and offer few course requirements and limited field experiences in mathematics (Horm, Hyson, & Winton, 2013). Because few faculty members are well prepared in ECME, the road begins with efforts to support their learning, from their graduate programs to ongoing faculty development, perhaps through professional associations such as the National Association for the Education of Young Children (NAEYC). Other chapters in this book offer helpful guidance on the adequacy issues as applied to early childhood mathematics, including descriptions of goals for teachers and of children's developmental trajectories. For ECME, there is an additional challenge of persuading early childhood teacher candidates that mathematics can and should be taught in the early years, which coincides with a need to increase teachers' self-efficacy and motivation to teach mathematics (Ginsburg, Lee, & Boyd, 2008; National Research Council [NRC], 2009). Fortunately, there is some evidence that repeated exposure to high-quality mathematics teaching can help improve prospective early childhood teachers' sense of confidence and competence as future teachers of mathematics (Rosenfeld, 2011).

In spite of a growing body of research, curriculum, and examples of promising pedagogies, the de-emphasis of mathematics in early childhood classrooms is in part created by the lack of attention it receives in teacher preparation programs. The lack of attention to ECME in teacher preparation is also influenced by the perceived lower priority that math has in programs for young children. A vicious cycle is thus set in motion, placing children's mathematical development at risk.

Second, teacher preparation programs must address the *authenticity challenge* by devising new and more thorough ways to expose teacher candidates to the practice of teaching and to children's thinking and learning, in mathematics as in other areas, and to offer candidates a series of opportunities to implement what they have learned, leading up to extended clinical placements in early childhood settings (National Council for Accreditation of Teacher Education [NCATE], 2010). Teacher preparation programs generally struggle to offer candidates enough opportunities to understand quality teaching, which can be difficult to represent, analyze, and scaffold in the college or university classroom. Even those higher education programs that are accredited or nationally recognized as meeting the NAEYC Standards for Early Childhood Professional Preparation often do not provide these opportunities in early childhood mathematics education (Hyson, 2008; Hyson & Woods, Chapter 2). Presently, at the baccalaureate level, more than 450 early childhood teacher education programs at NCATE-accredited institutions have received national recognition from NAEYC. NAEYC has also accredited 183 associate's degree programs at 156 institutions.

NEW APPROACHES TO HIGHER EDUCATION IN EARLY CHILDHOOD MATHEMATICS EDUCATION: MAKING THE MOST OF TECHNOLOGY

In light of these sobering realities, we turn to attempts to create useful models or resources for higher education in ECME, many of which involve technology in one way or another. New technologies offer promising approaches to old educational problems by expanding access to content, offering examples of practice, and creating opportunities to connect with a larger community of practitioners. This is true in general, but especially so in the arena of ECME, where many teacher education programs have had insufficient foundations in content, examples of good practices, and faculty expertise. As we will see, technology—especially rich video examples of mathematics learning in early childhood—can serve as a productive scaffold to strengthen these foundations.

The "ed tech" field is relatively broad, encompassing learning technologies for students of all ages; administrative applications for managing students, classrooms, and schools; hardware; networks; and more. The focus of this chapter is technologies designed to enhance the confidence and competence of early childhood teacher candidates enrolled in higher education and that of their instructors. We begin by highlighting the potential power of video technologies as tools for introducing education students to children's mathematical development and to high-quality mathematics teaching. With that background, we turn to some specific examples of how technology may be implemented and combined with other innovations in the service of improved early childhood teacher preparation.

Digital Video and Web-Based Video Distribution to Support Early Childhood Mathematics Education in Higher Education Programs

Making the Most of Video Analysis in Early Childhood Mathematics Education Teacher Preparation

Analysis of video segments, whether provided by faculty or produced by students, seems especially important in the field of ECME, where relatively few students are able to directly observe high-quality mathematics education or to reflect on and analyze their own competence in teaching mathematics to young children. Repeated viewings and discussion of both outstanding video exemplars and less-than-quality efforts will serve to expand future teachers' repertoires and promote greater interest and engagement in mathematics.

Video can be an essential tool for ECME teacher preparation by helping to address the challenges of exposing teacher candidates to young children's mathematical thinking, developmentally appropriate mathematics, and exemplary instruction and learning activities. Video can also be used poorly, however, as a sort of online television, which prompts a glazing-over effect for the viewer. This section advocates for an alternative approach: for video to be used as a manipulative to engage learners in active analysis and construction. Following this discussion, we provide some nuts-and-bolts tips for producing good-quality videos. We then turn to an in-depth example of the use of video analysis in a blended course on young children's mathematical thinking.

Shifting the Focus from Teacher to Child The majority of studies on video-based methods of teacher development concentrate on lesson analysis, including an extensive body of work in mathematics education. These studies range in focus from what teachers attend to in the videos (Star & Strickland, 2008), to teachers' development of observation and reasoning skills (Santagata, Zannoni, & Stigler, 2007), to cross-cultural comparisons of what teachers notice about the features of video lessons (Miller & Zhou, 2007). Studies specific to mathematics teaching and learning have demonstrated that video analysis has the power to increase teacher candidates' comprehension of subject matter and pedagogical content knowledge as well as to significantly enhance their perception of how specific teacher decisions can have an impact on learning outcomes for children (Sherin & van Es, 2005).

The ubiquity of video on the web, coupled with new online tools for selecting, annotating, and sharing video, has begun to change the landscape for teacher interaction with video (Rich & Hannafin, 2009). These tools not only enable teachers to more easily create, upload, and review videos of their own and others' classroom teaching, but they also encourage an approach to close viewing and analysis that is far easier than with nonweb video technologies. Not surprisingly, most of the research in this area focuses on the analysis of teaching practices (Calandra, Gurvitch, & Lund, 2008; Wang & Hartley, 2003; Yerrick, Ross, & Molebash, 2005). One exception to the rule is a study of teacher reflection using video that found that teachers became more sensitive to their own practice and, as a side effect, paid more attention to what children were doing (Rosaen, Lundeberg, Cooper, Fritzen, & Terpstra, 2008).

The effects for preservice teachers, including future early childhood educators, are likely to be similar as they develop an awareness of their practice and children's thinking. Learning to observe and understand children should be one of the primary goals of any analysis of classroom video. A focus on the content and methods of teaching is not enough; to be truly effective, prospective teachers need to learn more about the children they will teach, and videotaping and video analysis software should support investigations into children's thinking as a primary means of formative assessment and planning for further instruction. (See Chapter 3 for an in-depth analysis of the development of mathematical thinking of a single child, using multiple video examples.) Research on "teacher noticing" also converges with the mathematics reform literature (National Council of Teachers of Mathematics [NCTM], 2000; NRC, 2001), specifically teacher investigation of and building on children's thinking. Indeed, both tend to focus the interrelatedness of skills such as "attending to children's strategies, interpreting children's understandings, and deciding how to respond on the basis of children's understandings" (Jacobs, Lamb, & Philipp, 2009). Putting children at the center of the process should similarly change the emphasis of video-based instruction for teachers.

"Designed Video" as a Tool to Help Future Teachers Hone Their Perceptions Video alone is not sufficient. If we want to help future teachers develop an "enlightened eye" (Eisner, 1998) for observing children—to discriminate what matters from what does not—it is necessary to "educate their perception" by first giving them the tools to discern and differentiate before introducing them to new content or new experiences (Schwartz & Bransford, 1998, p. 335). This goal is consistent with Piaget's (1976) notion that "if they are not on the lookout for anything . . . they will never find anything" (p. 9). Schwartz and Hartman (2007) suggest a useful approach they call designed video, or the careful development or selection of video content with specific learning outcomes in mind—in our case, mathematics outcomes for early childhood teacher candidates. This process of selecting very specific purposes for the videos that are being developed can focus the learners on differing viewing goals, which might range from merely attending to or understanding something, to learning to perform a complex teaching task, each of which puts different demands on the instructor and the learner. To illustrate, multiple, well-designed video examples of young children engaged in a single task with an interviewer can be extremely helpful to show teacher candidates variations of understanding and performance as well as the evolution of thinking as children get older.

The close viewing process develops teacher candidates' enlightened eye for significant, interpretation-worthy moments, but ideally, candidates are also developing a sense that there is something worth discovering in the content. An additional challenge is to help teacher candidates develop sufficient curiosity and an orientation toward evidence that will motivate them to seek out this information within the video. Ideally, with enough practice, teacher candidates will transfer their observational skills to classroom assessment and continuously seek new evidence of learning to inform their pedagogical decision making. There is an interesting opportunity for research into whether early childhood teacher candidates can improve their classroom assessment skills through practice with video.

Strategies for Engaging Early Childhood Teacher Candidates in Video Analysis

In the college or university classroom, video can serve as raw material to be controlled, segmented, reorganized, reviewed, discussed, and debated as part of an active learning experience. Although instructors may have to adapt their teaching practices to learn how to use video as an object of analysis and the basis of class discussion, they may also use the opportunity to model close viewing for students and demonstrate how this approach can inform good thinking. "The instructor challenges them to *interpret*, to *cite empirical evidence* for the interpretations, to *justify their interpretations* in the face of counter-arguments, and in general to *go beyond mere opinions* or vague, ideological thinking" (Ginsburg, Cami, & Preston, 2009). Videos do not need to be lengthy. In fact, shorter segments may place greater emphasis on close viewing and demand less of limited attention spans. If teacher education students are encouraged to view a clip repeatedly, and their viewing is scaffolded through prompts and questions about the content, they can learn to look with increasing granularity. Withholding evidence by stopping the clip at significant moments may also help students develop an awareness of their own uncertainty, forcing them to make their best assessment using only the information that is immediately available to them and to plan to obtain more information. These repeated encounters with observable evidence constitute a form of practice in a controlled environment where the content is well known by the instructor and is viewed in common with other students studying the material.

Here are five strategies for engaging early childhood teacher candidates in video analysis to develop better understandings of children's mathematical thinking and appropriate curriculum and teaching strategies. These strategies can be used in face-to-face, blended, and online contexts and can be supported and, in part, replicated with other technologies, some of which are described later in this chapter. It is also worth noting that many of these strategies are also likely to be effective for in-service professional development, as discussed in other chapters of this book.

1. *Guided lessons.* Instructors select a video clip of a single episode—for example, an observation of free play among several children, a researcher interviewing a child, or a teacher's math lesson in a pre-K classroom—and create short, meaningful segments to be viewed in sequence by teacher candidates who must answer questions associated with each segment. For example, a question might ask "Why did the child answer that way?" and invite the viewer to supply evidence. After answering a question, the viewer might be asked to watch the segment again while paying attention to a specific detail or be given more information about the segment, a sampling of responses from other students, or even direct feedback on their own response, and then asked to reinterpret the content. The structure of the guided lesson (e.g., video segments, interpretive questions, and feedback or sample responses) represents a sort of lab experience for practicing observation, interpretation, and decision making. The guided lesson can be used to encourage teacher candidates to make connections to theoretical material, anticipate what might come next, propose the next question, or critique the teaching or assessment techniques of the adult in the video clip.

2. *Close viewing with targeted comparisons.* Instructors present students with a variety of examples of a single topic, such as children counting, creating patterns, or performing simple calculations. Students are asked to compare and contrast two clips in a short written commentary in which they cite specific segments, using the time code in the video to support a claim, much as they would when citing written text. Next, students are asked to study a response written by a peer and respond with a counterargument. In both the original response and the peer review, students are encouraged to adopt multiple perspectives and draw a clear distinction between what they know and what they do not know, with specific reference to observed evidence. As in the guided video lesson, students can be encouraged to suggest what they would do next to teach and/or assess the child. They might also critique the teaching or interviewing technique of the adult in the video.

Five Strategies to Engage Teacher Candidates in Close Video Analysis

1. Guided lessons
2. Close viewing with targeted comparisons
3. Community viewing and identification of evidence
4. Developmental trajectory analysis
5. Critique of one's own teaching performance

3. *Community viewing and identification of evidence.* Together in the classroom or online, students view a single extended clip and identify and discuss the critical moments, such as examples of children's thinking, opportunities to ask another question, or key strategies employed by the teacher or researcher. As a group activity, the exercise offers students an opportunity to share and debate their analyses, whether in real time or asynchronously online. Students can make connections to materials from the reading to demonstrate a deeper understanding of the content, argue a position about what the child knows or does not know, or suggest strategies for further teaching or assessment. An explicit learning objective of this activity is to help students develop an enlightened eye for what matters and transfer the skill of explaining and contextualizing the evidence from instructor to student.

4. *Developmental trajectory analysis.* Students review a single child's development over time—for example, using video interviews with a single child—to document the child's changes in mathematical thinking, increasing mastery of specific concepts, and ways of articulating what he or she knows. Students can complete detailed analyses of a specific year or trace a concept across multiple years and suggest strategies to facilitate learning toward anticipated skills. This method is modeled in Chapter 3.

5. *Critique of one's own teaching performance.* In this strategy, students record videos of their own teaching performance, upload the clip to the web, and conduct a close viewing and critical analysis of their work that involves citing specific evidence and creating a self-assessment using criteria established by the instructor. For their part, instructors can review the full clip or focus on the segments selected by the student. Students learn to recognize successful and less successful teaching behaviors, plan specific ways to improve their practice, and learn to use self-reflection as a tool for ongoing improvement. The process also yields a portfolio of examples of practice that students can use to demonstrate their own professional growth and from which to draw high-quality examples they can use during reviews or when seeking employment.

These are only a few of the many strategies that early childhood faculty can employ in ECME courses and field experiences. The design of these activities is based on pedagogical goals (such as those noted in Chapter 4), with the goals driving the use of video rather than being driven simply by whatever technology is available. With appropriate goals in mind, these strategies can leverage widely available technologies such as web-based video and communication tools to connect with teacher education students outside the physical classroom and to provide a virtual space where students can represent their own thinking, drawing upon cited evidence from video. Above all, the challenge is to avoid passive viewing and to encourage the active pursuit of evidence.

Nuts and Bolts of Finding, Making, Editing, and Annotating Videos

Implementing strategies for close video analysis of early childhood mathematics requires more than understanding the rationale, underlying concepts, and key strategies. It is true that, in theory, the advent of digital video and web-based distribution has greatly simplified the processes of recording, editing, and sharing video for both faculty and students. Within the past 10 years, a formerly cumbersome process involving videotape and unwieldy file sizes and types has become relatively easy, and even second nature, to many teacher candidates. In fact, many people now carry a powerful videocamera in their pockets in the form of a smartphone. Yet at the same time, to make the most of this potential, some practical nuts-and-bolts guidance is needed both for instructors and for their students. How can useful, high-quality video examples be identified or produced? Two main categories are addressed here: *high-quality exemplars* provided by faculty and *footage generated by teacher candidates*. Each can play a critical role in revealing the nuances of teacher practice, children's thinking, and their interrelationship.

Libraries of *high-quality exemplars* can be used by faculty as a multimedia companion for teaching about pedagogy, developmental psychology, and other content where visual evidence is a powerful complement to descriptive text. For ECME teacher preparation, these videos might focus on early childhood mathematics content, pedagogical methods, and assessment techniques, including the clinical interview. These videos can be faculty generated or licensed from vendors. Not all exemplars are equally valuable, however: A high-quality exemplar clearly illustrates a given mathematical concept in practice, ideally with a child revealing his or her thinking in discussion with an adult interviewer or a peer.

Video libraries are available as commercial products, including Alexander Street Press's database "Education in Video" (http://ediv.alexanderstreet.com/help/view/about_education_in_video), "Videatives: Video Clips for Early Education and Child Development" (http://www.videatives.com)—including some clips specific to mathematics—and as free public resources, such as "Teachers' Domain" from PBS and WGBH Educational Foundation (http://www.teachersdomain.org). Early childhood teacher education programs, whether housed in schools of education or other units, can also develop their own video libraries using carefully culled materials from local taping efforts such as the Video Interactions for Teaching and Learning (VITAL) project's early childhood mathematics collection, taped on location in New York City (http://ccnmtl.columbia.edu/vital/nsf). The VITAL project, to be discussed in detail later in this chapter, produced extensive video resources. Permission to use these resources is restricted, but others can produce similar materials in their own settings—with the benefit of making the videos locally relevant.

Whatever the source, it has become relatively inexpensive to create a collection of high-quality video content. To do so, however, requires a keen eye for teacher–child interactions; a decent videocamera, tripod or holder (especially if using a cell phone to record), and microphone; some knowledge of how to position the camera to avoid too-bright light sources and minimize extraneous noises and where to place the camera so that both the adult and child/children are in view. The Center for Advanced Study of Teaching and Learning (CASTL) at the University of Virginia is developing a manual for Head Start programs to assist them in making quality video records for analysis and assessment. Finally, plenty of time should be allotted for faculty to review the footage and make selections that are well matched to their goals for the teacher candidates who will view and analyze it.

Teacher Candidates' Videos of Their Own Practice An equally powerful application of digital video is *footage generated by teacher candidates* of their own practice, both for self-study and reflection and for mentor observation and feedback. In this case, speed trumps quality; the value of the footage is in teacher candidates' quick access to recordings of their own teacher–child interactions and lessons in order to compare their perceptions with others' feedback and to engage in a reflective process of identifying successes and areas for improvement. Teacher candidates' web-based study of their own video was pioneered by the MyTeachingPartner project (http://curry.virginia.edu/research/centers/castl/mtp) at the University of Virginia and (specific to mathematics teaching) by the VITAL project (http://ccnmtl.columbia.edu/vital/nsf) at Teachers College Columbia University. This kind of video footage is also very helpful for faculty supervisors, who can more readily observe and give feedback to student teachers in classroom settings, even if the supervisor is not actually present. The ability to provide this kind of quality supervision can greatly enhance the field component of teacher preparation programs, as recommended in recent national reports (NCATE, 2010). Lengthy segments are not necessary; on their own, teacher candidates may wish to review everything from their recent lessons, but they might share brief, selected clips with supervising faculty to save time and to focus on specific issues. Given the currently limited emphasis on ECME in teacher candidates' field experiences, specific assignments might be added, asking students to include mathematics lessons and activities in their video clips and then share these with classmates and faculty supervisors.

A new use of teacher candidate–generated videos is emerging with field testing and beginning state-level implementation of edTPA (http://edtpa.aacte.org), a video-based preservice assessment process developed in a partnership between Stanford University and the American Association of Colleges for Teacher Education. The process is intended to contribute to teacher licensure and program accreditation while ultimately improving candidates' readiness to teach effectively in their area of specialization. In addition, the videos are available to faculty and teacher candidates for reflection and improvement. If they are participating in edTPA, candidates record video of themselves during three to five classroom lessons related to a unit of instruction, along with submitting written lesson plans, samples of student work, and candidate commentaries. Videos are 15–20 minutes in length, which edTPA believes will be representative of longer segments of instruction. Currently, early childhood mathematics is not one of the areas of focus in edTPA; as of this writing, in the early childhood subject area, the videos must focus on literacy lessons only, although it is possible this could change in the future. Aligned with the Common Core standards, edTPA does offer a subject-specialized assessment in elementary grade mathematics, and the edTPA general elementary education subject area includes a math assessment task along with literacy.

Help with Selecting, Editing, and Annotating Video Materials Simple desktop editing tools such as iMovie for Mac and PowerDirector for PC can facilitate candidates' or faculty members' selection of key video segments. There are also a growing number of web-based tools for this purpose, including WeVideo (http://www.wevideo.com), Photobucket (http://www.photobucket.com), and

Kaltura (http://www.kaltura.com). Video hosting and sharing services include YouTube (http://www.youtube.com), Vimeo (http://www.vimeo.com), and Flickr (http://www.flickr.com), all of which offer varying video file-size and overall storage-capacity limits, as well as privacy controls.

Once online, video can also be analyzed using web-based platforms that support video clipping and annotations—that is, notes that can be typed or recorded, tied to specific time codes in the video, and shared with others. A few examples of these tools include Mediathread (http://ccnmtl.columbia.edu/mediathread), Project Pad (http://projectpad.northwestern.edu/ppad2), Video ANT (http://ant.umn.edu), and YouTube Video Annotations (http://www.youtube.com/t/annotations_about). The MediaThread platform can be used by faculty to construct a multimedia lesson of text, image, and video. See Figure 6.1.

One final note about the nuts and bolts: It is essential to obtain parental permission before making video recordings of children, regardless of one's intended use of the clips. Fortunately, there are many good models for permission letters, including edTPA's (http://assessment.pesb.wa.gov/assessments/edtpa).

The Video Interactions for Teaching and Learning Project: A Case Example of Using Video as a Pedagogical Tool in a Blended Course

Blended learning—the combination of online and face-to-face learning—can offer a sensible middle ground between virtual and traditional classrooms as well as opportunities for in-depth viewing and analysis of online video materials. When implemented well, blended learning redefines the roles of faculty and student and redistributes responsibility for learning. In a fully blended course, much of the content shifts online for students to explore and process, allowing the faculty member to repurpose face-to-face time to engage students directly,

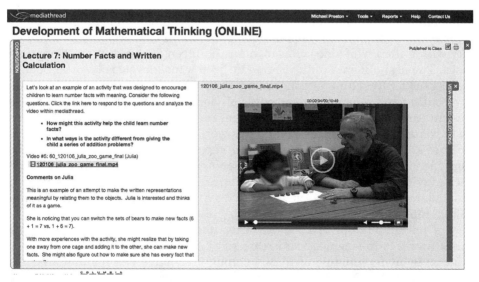

Figure 6.1. A lecture from the Development of Mathematical Thinking course at Teachers College, Columbia University, including a faculty-produced video exemplar illustrating number facts. (Screenshot courtesy of Columbia Center for New Media Teaching & Learning.)

employing a facilitative or coaching role. Students gain some decision-making power over content, pace, and time, and class time shifts to the more cognitively demanding modes of group work, discussion, and presentation. The key ingredients in a blended classroom are faculty feedback (constant formative assessment as students produce work) and student demonstration of mastery (work is not "done" until the instructor and the student agree that learning has occurred). A recent meta-analysis found blended instruction to be more effective than either conventional face-to-face or online instruction (U.S. Department of Education, 2010). Although the analysis focused on studies in K–12 settings, it is probable that similar patterns may be found in higher education.

To illustrate the inclusive use of online pedagogical tools in a blended course, we now take a closer look at a National Science Foundation (NSF)–supported model course in early childhood mathematics, the "Development of Mathematical Thinking" at Columbia University (http://ccnmtl.columbia.edu/vital/nsf; Ginsburg et al., 2009). As described earlier in this chapter and elsewhere (e.g., NRC, 2009; Horm et al., 2013), very few ECME courses are available to early childhood teacher candidates, and many existing courses fail to provide authentic examples of children's mathematical development and learning in the early years. Although the research literature on young children and mathematics is rich (NRC, 2009), early childhood teacher candidates are unlikely to be able to understand or apply this literature without having ready access to authentic examples from children and classrooms.

Before looking at the uses of technology in this course, the course's broader context should be understood. First, the course is—as other similar courses should be—more than a collection of decontextualized video examples. The course is grounded in developmental theory and in the research literature, within which the video material is placed and analyzed. Besides providing video examples, video lessons, and video analysis, the course engages students in readings, a real-world clinical interview activity, and an end-of-year interview-teach-interview-analyze-reflect project that bridges theory and practice.

Specific to the use of video technologies, the VITAL project provided faculty with a library of video clips of young children engaged in mathematical tasks. The assignments in the course utilized a tool developed by the Columbia Center for Teaching and Learning that allowed students, on their own, to go online and select short segments of video from the library, citing these clips in an interpretive written commentary completed outside of class. In these commentaries, teacher candidates were required to make claims about children's learning of mathematics and support their claims with evidence from the videos. They knew their commentaries would be reviewed and feedback provided online, and later their interpretations would be discussed and debated during face-to-face class time. See Figure 6.2 for a screenshot of the VITAL interface.

As just described, blended learning courses or modules that combine online and face-to-face methods create rich opportunities to enhance teacher candidates' understanding of ECME through reflective viewing and analysis of video examples. Nevertheless, for a number of reasons, preservice teacher educational programs may not be able to provide this kind of blended instruction in ECME. What are the alternatives?

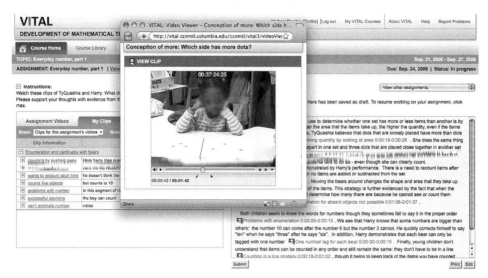

Figure 6.2. The VITAL "multimedia essay" with the students' collected video clips on the left side of the screen and a writing space incorporating text and video on the right. Students click or drag their video clips to add them to their essay. (Screenshot courtesy of Columbia Center for New Media Teaching & Learning.)

Alternatives to Blended Courses: Fully Online Instruction in Early Childhood Mathematics Education

Nearly every college and university offers at least a few courses online as an alternative or supplement to face-to-face courses. More recently, a few institutions have begun offering courses to large numbers of nonmatriculated students through massive open online courses, or MOOCs, which can serve thousands of students at a time. Specific to early childhood, in spring 2013, the University of Virginia launched the first MOOC designed for teachers and future teachers of young children, a 6-week course called "Effective Classroom Interactions: Supporting Young Children's Development." Although no specific plans for mathematics courses have been proposed, perhaps content-specific courses will soon follow, using this or other platforms. Obviously, it is easy to include ECME video examples in online courses, although the real-time debate and interaction embodied in the blended course just described would be missing.

Students who need the flexibility of time and distance can enroll in fully online degree programs offered by either virtual schools or traditional brick-and-mortar schools. Today, a student pursuing a degree in early childhood education can choose from online programs at virtual universities including Ashford, Capella, Concordia, Grand Canyon, and Walden, as well as from online degrees offered by traditional schools such as Northampton Community College in Bethlehem, Pennsylvania, and the University of Cincinnati in Ohio. Some of these programs may include strong preparation in ECME courses, but most do not. Nevertheless, the opportunity is there to use fully online instruction to create greater coverage of ECME content for a wider range of future teachers than is now the case.

The proliferation of these new courses and programs can dramatically increase access to educational opportunities, especially in a field such as early childhood where many current teachers are now attempting to upgrade their qualifications with little time and few financial resources. However, online instruction is not a magic solution: Many online courses are simply web-based versions of traditional lecture courses with little student engagement and support, potentially leading to poorer outcomes than face-to-face courses, particularly for lower-achieving students (Figlio, Rush, & Lin, 2010). These kinds of online courses are designed to inform rather than engage, offering students voluminous information but little opportunity to interact with an instructor or peers to deepen their learning, much less to practice a skill and receive feedback. Furthermore, many students do not necessarily know how to learn in an online context, which, when designed well, places additional demands on students to be self-motivated and self-regulated.

The best online programs focus on building community within each course, and some explicitly teach online learning skills to new students. Within a given course, faculty can use various techniques to scaffold students' self-management skills and require them to participate actively in discussion and the sharing of work. Many of these techniques are especially important for early childhood education students (Donohue, Fox, & Torrence, 2007), who are frequently both older than and less familiar with new technologies than other students in higher education.

Specific to ECME, all the advantages and disadvantages of fully online courses need to be considered when thinking about how best to prepare early childhood educators.

More Alternatives to Blended Courses: Embedding Video Elements into Traditional Early Childhood Mathematics Education Courses

At some institutions, faculty who are interested in using blended or fully online approaches to enhance ECME for their students may face policies and practical obstacles that extend beyond their own classrooms, such as when institutions grapple with the equivalence of face-to-face and online courses. But in smaller ways, individual faculty can certainly incorporate pedagogies informed by blended learning, even without shifting to fully blended or fully online approaches. For example, an instructor might introduce into an ECME syllabus specific activities that generate opportunities for students to select, explore, and analyze content, perhaps with some done on their own time, and devote a greater percentage of classroom time to group work or whole-class discussions in which students share what they have learned. An example similar to this requires no formal departmental or institutional commitment to changing the mode of instruction, but ideally, the institution would work with departments to provide training, help early childhood faculty understand platforms and tools, and provide additional course support when needed.

Supplementing the Use of Technology within Early Childhood Mathematics Education Courses: Mathematics Learning Everywhere

For current and future early childhood teachers and for the children they teach, opportunities to learn key mathematics concepts via technology have become ubiquitous and, in many cases, free. Indeed, it is hard to miss the recent abundance of apps and web sites offering math games, opportunities for practice, videos, curriculum, and even full courses. The Joan Ganz Cooney Center reports that 72% of the bestselling apps in the iTunes Store target preschool or elementary school children (Shuler, 2012). New resources appear on a seemingly daily basis, generating a crowded and often confusing landscape. They appear across all platforms, including the web, tablets, and smartphones. Some are specifically designed for children, but adults—including future early childhood educators and even some of their instructors—can often learn something from them, too. As emphasized in Chapter 2, there is considerable evidence that many teachers lack a firm understanding of mathematical content, which is an essential foundation for learning how to help young children learn and enjoy mathematics. The discussion here focuses on two types of digital math resources: tutorials and games.

Supporting Teacher Candidates' Math Knowledge Through Online Tutorials

The most prominent example of a math-related online tutorial system is the free, YouTube-based Khan Academy (http://www.khanacademy.org), started by Sal Khan, a former hedge fund manager who wanted to help his out-of-state cousins with their math homework. He uses simple illustration, screen capture, and voice-over tools to create what is now a widely recognized mode of online instruction: short video tutorials focusing on a single topic in mathematics and other related fields. Users range from school children to advanced university students to retirees, all whom are learning or refreshing their mathematical knowledge either independently or as part of a course. At the time of this writing, Khan's video on basic addition (http://www.youtube.com/watch?v=AuX7nPBqDts) has been viewed nearly 2 million times.

The Khan mathematics videos are different from the video clips of children and classrooms discussed earlier in this chapter. Nevertheless, even in their simple schematic form, they feature a number of design elements worth highlighting because of their relevance for those who may develop similar resources for early childhood teacher education. First, the clips are *compact and specific,* making them easily searchable and matched to learners' needs; second, they are *multimedia,* with tight linkages between visual and verbal representations; and third, they can be *self-paced,* meaning viewers can watch a clip (or segments of a clip) as many times as needed, and they can choose to watch only what they need. See the Basic Addition video as it appears on the Khan Academy web site in Figure 6.3.

As it has grown, Khan Academy has introduced more explicitly pedagogical tools and initiated partnerships with K–12 schools to help teachers use these resources in the classroom. Potential uses in higher education have also

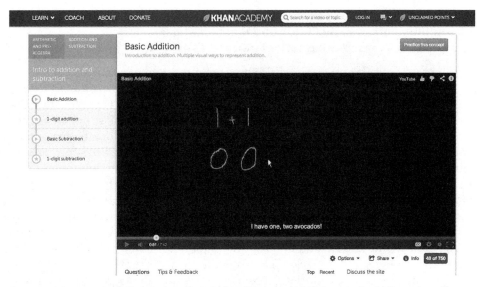

Figure 6.3. A moment in the Khan Academy "Basic Addition" lesson in which Khan uses a sketch of avocados to illustrate one-to-one correspondence. The Khan Academy math sequence covers arithmetic, algebra, geometry, trigonometry, calculus, statistics, and more at multiple levels of difficulty. (Screenshot courtesy of Khan Academy.)

been highlighted in Khan Academy's materials (https://www.khanacademy.org/coach-res/for-teachers/implementation-design/a/using-khan-academy-in-higher-education). Given the wide use of the Khan Academy tools, early childhood faculty may want to ensure that their students are familiar with these uses as well as consider whether online tutorials might help some early childhood students fill in gaps in their own mathematics background.

Again, a few design elements should be noted in the pedagogical tools that accompany the Khan Academy math tutorials. Short *assessments* at the conclusion of each video segment allow viewers to test their understanding and make an informed decision about whether to study the same content again or move along. Khan Academy's invitation for new people to contribute tutorials to the site suggests that *multiple perspectives* on a given topic can be represented. In summary, the Khan videos can be *studied by current and future teachers* for pedagogical purposes, *shown in higher education classrooms* (of both blended and traditional varieties), and even used to *support "flipped" classrooms* in which students study content away from the classroom, so class time can be devoted to working on assignments and projects with faculty support. Recently, the web site added data dashboards for students and teachers to track their progress and identify areas for further study.

Although not without some criticism (e.g., Khan's lessons typically lack explanation about the underlying rationale for algorithms, and the mathematics problems are not embedded in real-world situations; Schrag, 2013), Khan Academy videos can serve as worthwhile tools in higher education. That is, Khan Academy videos can be useful for teaching future early childhood educators some key mathematical concepts, and they also provide a useful model for teacher candidates to demonstrate their own mathematical knowledge by creating and sharing lessons of their own design.

Other Free Online Resources for Future Teachers to Build Children's Mathematical Understanding: Reviewing and Reflecting on Their Value

Although the video tutorial format offers maximum flexibility and choice for learners who may be struggling to master basic math concepts needed to support their work as future teachers, other resources may give teacher candidates—especially in the K–2 arena—ideas for how to support children's mathematical understanding. These resources offer personalization driven by adaptive technologies. For example, DreamBox (http://www.dreambox.com) and Time To Know (http://www.timetoknow.com) offer adaptive elementary mathematics instruction that uses performance data to determine whether children advance to new topics or continue to practice. Both products are designed to support children's blended learning in schools, and sometimes at home. Again, faculty may introduce their early childhood students to these resources and help students critique the resources' appropriateness for young children. As a bonus, students' independent exploration of the resources may help consolidate their own mathematical understanding.

Helping Future Teachers Consider the Value of Math Apps Children May Use at Home

At home, although the economic digital divide persists, demand is high among many preschool children for online games and apps. The increasing prevalence of cell phones, particularly smartphones, and the proliferation of cheaper computers make technology access less of an issue than in the past. Parents often share their smartphones with their young children, whether for popular games such as Angry Birds or Fruit Ninja, or perhaps a game with cleverly concealed educational content such as Park Math (http://www.duckduckmoose.com/educational-iphone-itouch-apps-for-kids/park-math), Team Umizoomi (http://www.nickjr.com/team-umizoomi), or Mathmateer (http://dan-russell-pinson.com/my-games). (Note that here and elsewhere in this chapter, the use of examples does not imply endorsement.)

Teacher education students should be introduced to these kinds of resources, discussing how they might be integrated into a comprehensive preschool mathematics curriculum, and what guidance might be given to parents about using such resources to promote math confidence and competence. Early childhood mathematics apps should focus on topics including shape recognition, number recognition, counting, more and less, sorting, and addition and subtraction. Besides emphasizing appropriate content, the best educational apps for pre-K children are fun and accessible, with interfaces that require minimal explanation or literacy skills; in contrast, too many other apps offer the equivalent of digitized workbooks that teach little besides rote memorization.

As educational tools, well-designed apps tap into young children's natural interest in using devices with screens. The fact that the apps often employ competitive game elements (such as points and badges, leveling up, and high scores both to show what the learner has accomplished and to hint at new content that must be "unlocked" by making progress) should prompt faculty to discuss with their students the pros and cons of these motivators. Issues about the appropriateness and effectiveness of these and other features of preschool math apps may be valuable topics for discussion in teacher education courses, in students' interviews with young children, or in field experiences in early childhood classrooms. Features might be examined in light of, for example, NAEYC's guidelines for developmentally appropriate practice or the NAEYC position statement on technology in early childhood education (http://www.naeyc.org/positionstatements).

Connecting Future Teachers with Early Childhood Mathematics Education Resources and with One Another

Again, technology makes it possible for future teachers of young children to stay professionally connected—important for many reasons, but especially in creating linkages to support students' confidence and competence in mathematics education. Similar to other aspects of early childhood teacher preparation, the ability to connect with fellow professionals has moved online, with positive implications for preservice professional development. There are increasing numbers of online communities and resource-sharing sites focusing on a range of topics including curriculum, standards, assessment, technology, and more. Some have an explicit focus on ECME, and others have that potential. A number of examples follow.

Just as faculty members encourage students to seek out professional resources from early childhood organizations such as NAEYC and the Division for Early Childhood of the Council for Exceptional Children, they can also help future teachers learn how to stay abreast of the latest technology resources. The tech-savvy teacher candidate should regularly check or subscribe to recommendation sites such as Common Sense Media (http://www.commonsensemedia.org), EdSurge (https://www.edsurge.com), and Edudemic (http://www.edudemic.com), again thinking about the relevance of these recommendations when teaching young children.

Teachers have shared their curriculum online for more than a decade. Recently, these efforts have acquired more official branding and gained traction through a nationwide effort to help teachers meet the implementation expectations of the Common Core State Standards. Two prominent examples include the NYC Department of Education's Common Core Library (http://schools.nyc.gov/Academics/CommonCoreLibrary), where teachers can search for Common Core–aligned tasks by grade level and discipline, and the American Federation of Teachers' Share My Lesson (http://www.sharemylesson.com), which includes a section devoted to mathematics for Grades K–2 (http://www.sharemylesson.com/early-elementary-math-teaching-resources) that lists thousands of resources, including "activities, worksheets, games, lesson plans, flashcards, puzzles, posters, assessments and other ideas." Because the Common Core focuses on only two subjects, one of which is mathematics, these kinds of resources, aligned with the standards, are valuable for preservice teacher education students as well as current teachers. Students should realize, however, that some of the activities may not be appropriate for children below kindergarten age and that all the activities should be examined to see how well they fit with generally accepted principles of developmentally appropriate, effective early childhood practice (a useful assignment, in fact). Besides these Common Core–related sites, there are also more open or community-driven organizations, such as CK-12 (http://www.ck12.org), which provides free access to texts, activities, and teaching materials (mostly for older students), and Curriki (http://www.curriki.org), which is an open platform for sharing and rating contributions of curriculum, including mathematics curriculum.

Specific to mathematics, the Erikson Institute's Early Mathematics Collaborative (http://earlymath.erikson.edu) offers online learning labs and an online library of video lessons, videos of children doing mathematics, and other math-related resources aligned with the Common Core. Online discussion boards support learning communities for early childhood professionals, including preservice teacher candidates.

Preservice teacher candidates need not rely on formal networks to connect with like-minded educators. Social media tools, including Facebook, Google Plus, and Twitter, enable current and future teachers to generate their own "personal learning networks" (PLNs) at a distance, with colleagues they may never have met in person. These networks allow for the sharing of resources and ideas, as well as a place to ask questions and request assistance. For example, the "#Kinderchat" community uses Twitter keywords (identified by the hashtag before each word) to organize a dialogue and sharing of early childhood education practices and resources. For practitioners interested in ECME, the search string "kinderchat math" (https://twitter.com/search?q=kinderchat+math) yields a list of links and discussions that updates almost daily. In an acknowledgment of the PLN phenomenon, the U.S. Department of Education recently established a site called "Connected Educators" (http://connectededucators.org) in order to create a hub for PLNs "to improve teacher and leader effectiveness, enhance student learning and increase productivity."

In the higher education setting, an ECME instructor might create a PLN for a class and require teacher candidates to locate and share resources that may be relevant to practical classroom problems related to weekly course topics—again, using class discussions to examine the resources in light of the kinds of practices recommended in other chapters in this book. Or faculty might require candidates to join an existing community, such as Kinderchat Math, and participate actively to acclimate them to the PLN world and establish connections that may be helpful in supporting them professionally. Such an assignment, with student assessments, also provides evidence of an early childhood teacher education program's ability to meet NAEYC accreditation standards in the area of "Becoming a Professional."

Introducing future teachers to these new approaches to building ECME-related learning communities requires time, professional development for faculty, and support from colleagues and outside partners with specific expertise. These communities—many of which blend online and face-to-face worlds—improve current and future professionals' ability to connect with one another outside their organizations and participate in a larger conversation about topics of common concern, including but not limited to mathematics education.

Building Early Childhood Faculty Capacity Through Online Communities and Other Supports

This chapter began by summarizing current critiques of teacher education, with special emphasis on the challenges of preparing teachers to help young children learn and enjoy mathematics. The chapter has described the promise of technology, and especially video-based and other online resources, to create conditions for preservice professional development that are more *adequate* and more *authentic* than what have been available in the past.

In creating these conditions, faculty in early childhood teacher education-programs are key. But it is not easy for faculty members to implement these new approaches, which require more than adopting some new technological gimmicks. Faculty need to begin by ensuring that their own theoretical and research foundations—as well as their interest in and commitment to early childhood mathematics—are adequate to the task. This is true whether an instructor is a specialist in ECME or is integrating math education content into courses in child development, curriculum, teaching strategies, or early childhood assessment. If faculty have this kind of foundation in theory and research, they are well positioned to critically explore the resources described in this chapter and to consider how best to include them in their teaching.

Faculty who undertake to use video resources and other technologies to enrich their students' preparation in ECME (whether through blended, fully online, or traditional approaches) should first commit to identifying valuable learning outcomes and then work backward to ensure that they are using technology to achieve something purposeful. In Chapter 4, Juanita Copley provides an excellent guide to worthwhile ECME goals for teachers; these can guide teacher educators' decisions as well. Next, faculty members should select and try out a new technological tool on their own or with colleagues to test its features with students' perspectives in mind and to get comfortable enough to be able to demonstrate it for their students. Finally, they should introduce the new technology to their students along with clear expectations for the frequency and quality of work students will produce.

It is easier to make these recommendations than to implement them in higher education. As noted earlier, even highly motivated faculty face a number of difficulties. Access to resources and opportunities for faculty professional development in ECME have been limited. Professional isolation adds further barriers when considering and adopting new practices. Several approaches may help college and university instructors move ahead.

Virtual Learning Communities for Faculty

Some of the opportunities for math-related professional networking that were recommended for teacher candidates may support faculty as well. Virtual learning communities, whether formal or informal (the PLNs previously described), can serve many functions; however, they may be especially helpful in building faculty members' capacity and confidence in ECME, where many instructors have acknowledged gaps in their own preparation (Hyson, 2008).

A future source of relationships with other instructors interested in ECME may be a web resource called Early Childhood Faculty Connections (http://www.facultyconnections.org), now in prototype form. When fully functioning, the site will serve as a resource for the higher education community, with opportunities to share syllabi, assignments, and other pedagogical tools; links to new research; mentoring and collaboration information; and connections to projects and professional organizations. Faculty members with special interest in ECME would be able to use the site to connect with others, share course syllabi and assignments, and learn from colleagues.

CONCLUSION

These practices and tools do not portend a world in which machines replace human instructors. Rather, they establish the conditions for instructors to focus their energies in more meaningful ways, such as supporting teacher candidates' thinking about children's mathematical development and about effective ECME practices, using the support of technology. The best educational innovations help instructors do their work better while also enhancing relationships between instructors and learners. Feedback and support for a developing teacher cannot be found in Google search results. This chapter advocates for technology that increases student engagement and critical thinking, shifts the cognitive "heavy lifting" from faculty instructor to teacher education student, expands access to mathematics skill building activities (for both adults and young children) and makes them fun and engaging for all, and provides opportunities to understand expert practice and develop one's own confidence and competence with skilled guidance and support. Finally, we must emphasize the need for high-quality research on every aspect of preservice teacher preparation in early childhood mathematics, critically examining the impact of key features of that preparation—including but not limited to technological tools—on teachers' practices and on young children's mathematical competence.

FOR REFLECTION AND ACTION

1. Preston outlines a set of challenges faced by all teacher preparation programs today—not only in preparing early childhood educators to teach mathematics. From your experiences in such programs or from discussing this with others, to what extent do you encounter these challenges? What, in your view, are the practical effects of inadequate preparation of teachers within higher education?
2. Specific to learning how to teach mathematics to young children, what were your experiences or what are the expectations within your state's higher education institutions and its teacher certification system? From your experience and from exploring other chapters in this book, what appear to be the problems with current content and methods?
3. Preston emphasizes the role of technology in providing preservice teachers with what they need to see and experience as they learn how to be effective in implementing

mathematics instruction. In your setting, can you inventory the kinds of technology supports that are currently available or that might be used for this purpose? If you are a student or faculty member, you might think about your institution; if you are involved in policy, you might think about state-level technology initiatives within the higher education system.

4. The chapter makes it easy for you to explore some of the innovative technology options that can enhance the preparation of college and university students in ECME. On your own or with colleagues, take time to explore a number of these from the book's downloadable content. How might you or your institution use or adapt these kinds of resources?

5. Many other recommendations are woven throughout Preston's chapter. If you are a faculty member or administrator, you might use the "Early Childhood Math Education Action Planning Tool" found in the downloadable content to note specific examples that you will implement in your college or university teaching or that you will recommend to faculty for adoption. Along the same lines, be sure to download the examples of course syllabi and other materials; again, these may be adapted for your needs and those of your students.

REFERENCES

Calandra, B., Gurvitch, R., & Lund, L. (2008). An exploratory study of digital video as a tool for teacher preparation. *Journal of Technology and Teacher Education, 16*(2), 137–153.

Darling-Hammond, L. (2011, March 16). Teacher preparation is essential to TFA's future. *Education Week*. Retrieved from http://www.edweek.org/ew/articles/2011/03/16/24darling-hammond.h30.html

Donohue, C., Fox, S., & Torrence, D. (2007). Early childhood educators as eLearners: Engaging approaches to teaching and learning online. *Young Children, 62*(4), 34–40.

Eisner, E.W. (1998). *The enlightened eye: Qualitative inquiry and the enhancement of educational practice.* Upper Saddle River, NJ: Merrill.

Figlio, D., Rush, M., & Yin, L. (2010). *Is it live or is it Internet? Experimental estimates of the effects of online instruction on student learning* (National Bureau of Economic Research Working Paper No. 16089). Retrieved from http://www.nber.org/papers/w16089

Ginsburg, H.P., Cami, A., & Preston, M.D. (2009). Inquiry practices: How can they be taught well? In N. Lyons (Ed.), *Handbook of reflection and reflective inquiry: Mapping a way of knowing for professional reflective inquiry* (pp. 453–472). New York, NY: Springer.

Ginsburg, H.P., Lee, J.S., & Boyd, J. (2008). Mathematics education for young children: What it is and how to promote it. *Society for Research in Child Development Social Policy Report, 22*(1), 3–22.

Horm, D.M., Hyson, M., & Winton, P.J. (2013). Research on early childhood teacher education: Evidence from three domains and recommendations for moving forward. *Journal of Early Childhood Teacher Education, 34*(1), 95–112.

Hyson, M. (2008). *Preparing teachers to promote young children's mathematical competence.* Paper commissioned by the Committee on Early Childhood Mathematics. Washington, DC: National Research Council.

Hyson, M., Horm, D.M., & Winton, P.J. (2012). Higher education for early childhood educators and outcomes for young children: Pathways toward greater effectiveness. In R. Pianta, L. Justice, S. Barnett, & S. Sheridan (Eds.), *Handbook of early education* (pp. 553–583). New York, NY: Guilford Press.

Jacobs, V.R., Lamb, L.L.C., & Philipp, R.A. (2009). Professional noticing of children's mathematical thinking. *Journal for Research in Mathematics Education, 41*(2), 169–202.

Mashburn, A.J., Pianta, R.C., Hamre, B., Downer, J., Barbarin, O., Bryant, D., . . . Howes, C. (2008). Measures of classroom quality in prekindergarten and children's development of academic, language, and social skills. *Child Development, 79*(3), 732–749.

Miller, K.F., & Zhou, X. (2007). Learning from classroom video: What makes it compelling and what makes it hard. In R. Goldman, R. Pea, B. Barron, & S. Derry (Eds.), *Video research in the learning sciences* (pp. 321–334). Mahwah, NJ: Lawrence Erlbaum Associates.

National Council for Accreditation of Teacher Education. (2010). *Transforming teacher education through clinical practice: A national strategy to prepare effective teachers.* Washington, DC: Author.

National Council of Teachers of Mathematics. (2000). *Principles and standards for school mathematics.* Reston, VA: Author.

National Research Council. (2001). *Adding it up: Helping children learn mathematics.* Washington, DC: National Academies Press.

National Research Council. (2009). *Mathematics learning in early childhood: Paths toward excellence and equity.* Committee on Early Childhood Mathematics, C.T. Cross, T.A. Woods, & H. Schweingruber (Eds.). Center for Education, Division of Behavioral and Social Sciences and Education. Washington, DC: National Academies Press.

NICHD Early Child Care Research Network. (2000). Characteristics and quality of child care for toddlers and preschoolers. *Applied Developmental Science, 4*(3), 116–135.

Piaget, J. (1976). *The child's conception of the world.* Totowa, NJ: Littlefield, Adams.

Rich, P., & Hannafin, M. (2009). Video annotation tools: Technologies to scaffold, structure, and transform teacher reflection. *Journal of Teacher Education, 60*(1), 52–67.

Rosaen, C.L., Lundeberg, M., Cooper, M., Fritzen, A., & Terpstra, M. (2008). Noticing noticing: How does investigation of video records of practice change how teachers reflect on their experience? *Journal of Teacher Education, 59*(4), 347–360.

Rosenfeld, D. (2011). *Fostering confidence and competence in early childhood mathematics teachers.* (Unpublished doctoral dissertation). Columbia University, New York, NY. Retrieved from http://hdl.handle.net/10022/AC:P:12157

Santagata, R., Zannoni, C., & Stigler, J.W. (2007). The role of lesson analysis in preservice teacher education: An empirical investigation of teacher learning from a virtual video-based field experience. *Journal of Mathematics Teacher Education, 10*(2), 123–140.

Schrag, F. (2013). Is this the education revolution we've been waiting for? An essay review of the one world school house. *Education Review, 16*(7). Retrieved from http://www.edrev.info/essays/v16n7.pdf

Schwartz, D.L., & Bransford, J.D. (1998). A time for telling. *Cognition and Instruction, 16*(4), 475–422.

Schwartz, D.L., & Hartman, K. (2007). It is not television anymore: Designing digital video for learning and assessment. In R.P.R. Goldman, B. Barron, & S.J. Derry (Eds.), *Video research in the learning sciences* (pp. 335–348). Mahwah, NJ: Laurence Erlbaum Associates.

Sherin, M.G., & van Es, E.A. (2005). Using video to support teachers' ability to notice classroom interactions. *Journal of Technology and Teacher Education, 13*(3), 475–491.

Shuler, C. (2012). *iLearn II: An analysis of the education category on Apple's App Store.* New York, NY: Joan Ganz Cooney Center. Retrieved from http://www.joanganzcooneycenter.org/wp-content/uploads/2012/01/ilearnii.pdf

Star, J.R., & Strickland, S.K. (2008). Learning to observe: Using video to improve preservice mathematics teachers' ability to notice. *Journal of Mathematics Teacher Education, 11*(2), 107–125.

U.S. Department of Education, Office of Planning, Evaluation, and Policy Development. (2010). *Evaluation of evidence-based practices in online learning: A meta-analysis and review of online learning studies.* Washington, DC: Author.

Wang, J., & Hartley, K. (2003). Video technology as a support for teacher education reform. *Journal of Technology and Teacher Education, 11*(1), 105–138.

Yerrick, R., Ross, D., & Molebash, P. (2005). Too close for comfort: Real-time science teaching reflections using digital video. *Journal of Science Education, 16*(4), 351–375.

7

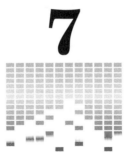

Promising Approaches to Early Childhood Mathematics Education Professional Development in In-Service Settings

Kimberly Brenneman

Earlier chapters detail the current state of early childhood mathematics education (ECME) with regard to teacher preparation and practice. These discussions inform this chapter's review of in-service professional development (PD) for ECME. Hyson and Woods' (Chapter 2) discussion of the characteristics of the early childhood teaching work force and the extent to which the profession, as a whole, is prepared—or not—to support young children as mathematics learners provides a rationale for the importance of in-service professional development for ECME. Due to great variability in the educational background of those who teach young children and continuing inadequacies in preservice training programs, even those educators who have college degrees are likely have little or no preparation to teach mathematics to young children. Zaslow's discussion of the general features of successful PD is relevant for a review effective in-service PD models, programs, and strategies for ECME (Chapter 5; see also Vick Whittaker and Hamre's Chapter 8 on evaluation of ECME PD), and some of the challenges to effective practices for in-service education, and potential solutions to them, parallel those identified by Preston (Chapter 6) for preservice education.

In what follows, I briefly review features of effective in-service PD for ECME, drawing on both the general PD literature and the much smaller, math-specific literature. Three research-based PD approaches that embody many of these characteristics are highlighted. Next, I present factors that administrators and others who select PD approaches for educational organizations should consider as they search for ways to improve mathematics teaching in their particular early childhood settings, as well as some considerations for PD designers and providers. Challenges to successful PD implementation are also described. Potential solutions to these challenges are presented, with the discussion drawing on existing research studies, anecdotal evidence from PD providers and researchers, and promising approaches to ECME PD that are currently being designed and implemented on a small scale. Finally, I present a wish list for research, policy, and practice in ECME PD. The list grows from issues that the chapter raises about the challenges of developing, researching, choosing, and implementing high-quality ECME PD of the kind that leads to positive change in teaching and, ultimately, to young children's learning of mathematics.

CHARACTERISTICS OF EFFECTIVE PROFESSIONAL DEVELOPMENT FOR EARLY CHILDHOOD MATHEMATICS EDUCATION

What constitutes effective professional development depends on one's goal (see Copley, Chapter 4). PD goals for practicing educators might include earning a degree, learning to implement a new curriculum, or increasing knowledge of a particular content area and how to support its learning and development. Scope can be relatively wide or narrow, perhaps encompassing a full mathematics curriculum or focusing on children's learning of a specific numerical operation. The delivery of training and PD varies and can include some combination of college coursework, teacher work groups or communities of practice, individualized coaching, web-based support materials, workshop series, and stand-alone workshops. PD delivery can occur in-person, via technology, or through a blended approach of the two. The dosage levels (how often PD is provided, for how long, and on what schedule) and the degree of individualization for the adult learner vary greatly among, and within, these types of PD.

As Zaslow (Chapter 5) details, in general, successful PD approaches are those that have specific and clear objectives, focus on long-lasting changes in practice, last long enough to fully explore the content being conveyed, center on content knowledge and link this explicitly to practice, are aligned with standards of practice, involve educators in active learning, involve joint participation from educators at the same site, and are part of a set of coherent PD experiences (see also Birman, Disimone, Porter, & Garet, 2000; Zaslow, Tout, Halle, Whittaker, & Lavelle, 2010a, 2010b). The authors of a 2009 National Research Council (NRC) report on early math reviewed both the general PD literature for early childhood and elementary school and the ECME literature to identify the characteristics of successful PD for ECME specifically. They concluded that a PD approach is more likely to lead to positive change in the attitudes, skills, and knowledge that affect practice and children's learning if it has the following features:

- Is standards based, ongoing, and embedded in the job so it is practical and concrete

- Involves a core group of colleagues who have time to work and learn together

- Includes stable, high-quality PD sources that allow for mentoring, experimentation in teaching, and reflection on practice
- Is based on a theoretical foundation and structured as a coherent system
- Responds to an individual educator's unique background, experiences, teaching context, and role as an educator
- Addresses mathematics knowledge as well as teaching strategies
- Is grounded in specific curricular materials that focus on children's thinking, learning, and mathematical development
- Matches the intensity and duration of the PD intervention to the content and scope of the objectives

For educators, administrators, and policy makers searching for PD programs that are likely to be effective, this list provides general guidelines to evaluate various ECME PD approaches for potential implementation. In the next section, I highlight three ECME PD models that exemplify many characteristics of high-quality PD. The models vary in a number of ways, including specific content; whether a curriculum is linked to the PD offerings; the ways that workshops and individualized supports are combined; whether PD is provided in-person, through technology, or in some combination; the grades targeted; and so forth. Each, however, has been evaluated for effectiveness for improving early childhood mathematics teaching and children's learning, and each has shown positive results that support use in early childhood settings, at least under circumstances similar to those in which the models were tested. Given initial successes, these approaches to ECME PD continue to be developed, expanded, and tested in new ways—with more classrooms, with different age groups, or in combination with PD for other curricular areas—with the goal of increasing our knowledge about the conditions under which each model is effective, thus further developing the toolkit of proven PD approaches available for adoption by educators.

THREE RESEARCH-BASED MODELS FOR EARLY CHILDHOOD MATHEMATICS EDUCATION PROFESSIONAL DEVELOPMENT[1]

Technology-Enhanced, Research-Based Instruction, Assessment, and Professional Development[2]

Features of the Approach

Technology-enhanced, Research-based, Instruction, Assessment, and professional Development (TRIAD) is a general approach to implementing and studying curriculum-based PD approaches as they are scaled up or expanded. Julie Sarama and Douglas Clements have applied the TRIAD model to their own work that targets prekindergarten educators who teach young children in low-resource communities. In some cases, follow-through PD is provided to K–1 teachers so that they can build on the teaching approach used in pre-K, with the expectation that a consistent approach across grades will lead to increased math achievement for children.

TRIAD aims to improve children's mathematics readiness by supporting teachers to provide high-quality instruction using the Building Blocks (BB) math curriculum for prekindergarten. (Supplemental materials to enrich existing curricula are available for K–6.) BB is based on standards for early math learning and on research-based descriptions of the sequences through which children's learning and development progresses. These learning trajectories form the core of the curriculum and are the focus of the PD experiences offered to teachers.

In addition to providing a curriculum and supporting materials, main components of the TRIAD PD model include a 4- to 5-day-long summer institute with a 2-day follow-up workshop, monthly classes, and classroom-based coaching. In addition, TRIAD includes a web application designed to support teachers as they learn to teach using learning trajectories. (A demonstration can be viewed on the TRIAD web site: http://gse.buffalo.edu/org/triad/tbb/index.asp?local=parent.)

Research Results

Studies of BB and, by extension, the PD approach associated with it have found positive effects on the classroom instructional environment for math and on children's learning (Clements & Sarama, 2008; Sarama, Clements, Starkey, Klein, & Wakeley, 2008). Compared with classrooms in which no changes were introduced and to classrooms that implemented another curriculum, teachers trained to use BB engaged children in a greater number of different math activities, in more reflective conversations about math, and in more time with computer-based math learning activities.

Children in BB classrooms learned more math as well. They scored higher on a comprehensive mathematics assessment than children in either the business-as-usual or the alternative math curriculum groups. Specific skills that were more advanced in the BB group than in the business-as-usual (and often the alternative math curriculum comparison) group included verbal counting, number recognition, simple arithmetic tasks, shape recognition, and shape composition, among others.

Positive effects for oral language were also found, with BB/TRIAD children showing better ability to recall key vocabulary from a story, more use of advanced grammatical structures when retelling, more willingness to produce a story independently, and improved ability to reason inferentially about the story's content. The effects were greatest in classrooms with a high-quality mathematics environment (measured with an observational tool designed to assess the quality and quantity of materials and supportive instructional activities for mathematics) and in which teachers engaged children with more frequent, even very short, mathematical activities (Sarama, Lange, Clements, & Wolfe, 2012).

Summary and Next Steps

Although it is not yet possible to separate the effects of the BB curriculum from the effects of the TRIAD approach to PD (see also the discussion in Chapter 8), these together form a powerful package that has been subjected to rigorous testing and has been shown to be effective under a variety of implementation conditions, including classrooms in different geographic locations using different comprehensive curricula. Some characteristics of the approach that likely influence its success are the tight link to a math curriculum that provides teachers with specific activities and instructions for implementing them, the coherent PD system that blends intensive workshops with individualized coaching and technology-enhanced support, and the strong theoretical grounding and foundation in how children learn mathematics.

The University of Denver team is currently involved with collaborative projects that test whether their approach for math and math PD can be successfully combined with a curriculum and PD approach that targets self-regulation (Clements, Sarama, Layzer et al., 2012) and with an interdisciplinary curriculum that includes strong supports for literacy, science, and social-emotional development (Clements, Sarama, Duke, Hemmeter, & Brenneman, 2012). The overarching goal of such work is to determine whether teaching practice and children's learning can be enhanced in multiple areas without diminishing the gains for math or other areas that are found when each is the sole focus of curriculum and PD.

MyTeachingPartner-Math/Science[3]

Features of the Approach

MyTeachingPartner-Math/Science (MTP-M/S) consists of a curriculum and related PD designed for preschool teachers and is implemented in state-funded prekindergarten programs targeting 4- and 5-year-olds at risk for lack of school readiness (Lee, Kinzie, & Whittaker, 2012). MTP-M/S supports teachers in their practice for math and science as well as language and literacy related to these. Coupling math and science seems natural. Math is sometimes called "the language of science," and there is clear overlap among the domains (Epstein, 2014), especially with regard to patterns, measurement, and data analysis. Meaningful science experiences often make use of mathematics, and science provides a purpose for children to use mathematical reasoning and skills.

The MTP-M/S PD model makes significant use of web-based video resources for teacher PD and learning. Blended with 8.5 days of face-to-face PD workshops to build skills, the online supports offer teachers the opportunity to work on PD when it is convenient and when a specific need arises, and teachers individualize their professional development by choosing which online resources to access (Lee et al., 2012). These resources include more than 100 brief video clips that model effective instructional practice using the MTP-M/S curriculum, in conjunction with a larger Quality Teaching Video Library, which includes teaching examples across domains (Pianta, La Paro, & Hamre, 2006).

Research Results

As an illustration of the complete PD approach, consider specific experiences that focused on helping teachers incorporate more higher-order, open-ended questioning into their practice. Teachers participated in two face-to-face questioning sessions as part of larger workshops. They also had access to teaching tips for questioning, demonstration videos that provided information about different types of questions and strategies for using open-ended questions, and examples of teachers using open-ended questioning strategies during classroom activities (Lee et al., 2012). Participants made two videos of themselves teaching, before and after the questioning PD. They engaged in focused reflection on their videos, noting the number of open versus closed questions asked and identifying the questioning strategies used.

Compared to teachers who implemented the MTP-M/S curriculum but did not receive the PD supports, participating educators asked more open-ended questions in the classroom; other sorts of questions did not differ among groups. Participants who made more use of the web-based, question-specific supports tended to ask more open-ended questions in the classroom. Learning gains were not limited to teachers, however, with children showing more lexical and sentence structure complexity than those in the comparison group (Lee et al., 2012). Because the type of teaching supports provided was varied in this study, it is possible to answer questions about the "active ingredients" in this PD approach, specifically about the value added by incorporating the web-based supports (for more details on MTP-M/S, see Chapter 8; authors Vick Whittaker and Hamre are codevelopers of the program).

Summary and Next Steps

Use of the MTP-M/S curriculum and PD has positive effects for teaching and learning in areas related to mathematics and science. The combined approach allows teachers to improve instructional offerings in multiple school readiness domains that have been shown to be associated with later achievement in math, science, and reading (Duncan et al., 2007; Grissmer, Grimm, Aiyer, Murrah, & Steele, 2010). Some characteristics of the PD that likely contribute to its success are its foundation in specific curricular materials, its focus on demonstrating practical teaching strategies, and individualized supports that respond to a teacher's self-identified needs.

On the strength of the initial research results, the MTP-M/S project recently received funding from the Institute of Education Sciences for efficacy trials to determine the extent to which the MTP-M/S program has an impact on the quality of teacher–student interactions and children's knowledge and skills in math and science (Center for Advanced Study of Teaching and Learning, n.d.).

Early Math Collaborative[4]

Features of the Approach

The Early Mathematics Education Project was launched in 2007 to provide in-service PD in early mathematics for 300 pre-K and kindergarten teachers in Chicago public schools. Now known as the Early Math Collaborative (EMC), the program lasts 1 year and involves five workshops (one every other month), on-site coaching, and video-based materials to foster teachers' reflective practice. Its goals are to increase teachers' math content knowledge and confidence in math teaching abilities (McCray & Chen, 2011).

Several key features characterize the program. First, it incorporates storybooks as a way to build on the strengths of teachers, who tend to be more comfortable with literacy than math. PD focuses on engaging teachers with the big ideas of mathematics (modules include sets, pattern and regularity, number, counting, operations, linear measurement, other measurement, shape, spatial thinking, and data analysis). Knowledge is expanded, and classroom applications are explored through a combination of active adult learning tasks, lecture materials, lesson plans that enable teachers to think about ways to introduce concepts to their children, and activity extension ideas. Videos are used to show classroom lessons and to focus on children's learning of particular concepts. Teachers also are provided with explicit guidance in the use of math language, which has been found to positively affect students' mathematical knowledge (Klibanoff, Levine, Huttenlocher, Vasileyva, & Hedges, 2006) but is presently lacking in many preschool classrooms (e.g., Rudd, Lambert, Satterwhite, & Zaier, 2008). This kind of language involves labeling mathematical situations and asking questions that require mathematical thinking—for example, noting that there are four children so four plates are needed or asking children to figure out how many plates are needed so each child has one. The project also involves meetings with principals and assistant principals across the participating schools so that these administration leaders can learn more about the project and what teachers are doing and learning (Early Mathematics Education Project Update, November 2012, provided by Jennifer McCray, December 30, 2012).

Research Results

Teacher participation in EMC is linked to positive child outcomes, with preschool students outperforming matched comparisons on the Applied Problems Subtest of the Woodcock Johnson-III (Woodcock, McGrew, & Mather, 2001). They showed almost 3 months' worth of additional learning, and those children who were already below national norms showed gains of 5 months of additional learning over and above that shown by comparison children. The intervention group also showed reliably greater growth as assessed by a comprehensive mathematics instrument, Klein & Chen's (2006) Child Math Assessment (CMA; McCray & Chen, 2011).

Summary and Next Steps

The EMC project exemplifies a number of the characteristics of high-quality ECME PD. These include having a coherent system of PD (i.e., workshops, coaching, and teacher work groups), being grounded in standards for early math learning and teaching, and providing individualized, classroom-based supports for teachers. The inclusion of leadership meetings with administrators is another feature of the approach that could support better mathematics teaching and stronger effects (discussed later in the chapter).

On the strength of early results in pre-K and kindergarten, the project team received an Investing in Innovation (i3) grant from the U.S. Department of Education to expand the approach to Grade 3, with more than 120 pre-K–3 teachers in eight schools in the Chicago Public School District. Teachers participate in workshops ("learning labs"), receive classroom-based coaching in math instructional practices, and engage in monthly meetings with other grade-level teachers at their school. Classroom practice is filmed as an object for reflection during coaching meetings, and teachers have access to online training videos and other resources to improve mathematics teaching. The Erikson group plans to adapt and test their approach for preservice training in ECME (J. McCray, personal communication, December 30, 2012).

Although these three ECME PD models are not the only ones that have been researched, each has proven effective under particular circumstances and is being tested under different, expanded conditions, meaning that we will soon know more about the unique benefits and potential limitations of each approach. Many PD programs exist, and for those who are charged with choosing ECME PD, being a careful consumer requires questioning the evidence of effectiveness of *any* candidate program. Although one might choose an approach that is under development (perhaps to participate in a research and development study), rather than completely researched, it remains critical to assess whether the PD program being considered has the characteristics of high-quality ECME PD and was designed using solid research evidence about child and adult learning.

PD is not a one-size-fits-all endeavor, though, so a thoughtful accounting of your specific goals for PD and of the unique features and potential challenges of your educational setting is essential for finding the match with the greatest potential for success. In the next section, I review some challenges or questions to consider when choosing and implementing a PD model or program and identify potential solutions from existing PD approaches and from "in the pipeline" models that are being developed by researchers and providers of ECME PD.[5] I provide detailed descriptions of two approaches that show promise for addressing one specific challenge: meeting the needs of children from diverse linguistic and cultural backgrounds. Along with TRIAD, MTP-M/S, and EMC, some of the in-the-pipeline approaches (see Table 7.1) have the potential to be added to the toolkit of ECME PD available to administrators, education policy makers, and teachers when they search for in-service training that is likely to be effective for their unique goals and circumstances.

CONSIDERATIONS FOR CHOOSING AND IMPLEMENTING EARLY CHILDHOOD MATHEMATICS EDUCATION PROFESSIONAL DEVELOPMENT

Research suggests that successful PD meets many of the previously listed characteristics, but producers and consumers of this kind of PD know that it is neither simple nor inexpensive to implement. It requires a sustained commitment on the part of teachers, administrators, and providers. Specific features of educational settings and systems have an impact on (and often impede) the implementation of the kinds of PD that are most likely to bring about positive change in ECME practice and learning outcomes.

Time for Long-Term, Sustained Professional Development

Time has an impact on ECME PD in a number of ways. One concerns the sheer amount of time required to effect change in teacher practice—that is,

Table 7.1. Five approaches to early childhood mathematics education in-service professional development

Project title	Lead researcher and institution	Selected key features	Development status
Technology-enhanced, Research-based, Instruction, Assessment, and Professional Development (TRIAD)	Julie Sarama and Douglas Clements—University at Buffalo and University of Denver	• Targets comprehensive mathematics • Provides prekindergarten curriculum and curriculum-based professional development for teachers • Has extensions available for K–6 • Uses blended approach of web-based and in-person workshop and coaching supports • Was developed and tested in varied auspices serving children in low-resource communities	Positive results for classroom quality, math learning, and oral language from a large-scale efficacy study Math curriculum and professional development supports being paired with nonmath curricular areas in new research and development studies To learn more, visit http://gse.buffalo.edu/org/triad/tbb/index.asp?local=parent
MyTeachingPartner-Math/Science (MTP-M/S)	Mabel Kinzie—University of Virginia, Center for Advanced Study of Teaching and Learning	• Targets both math and science • Provides prekindergarten curriculum and curriculum-based professional development for teachers • Uses blended approach of web-based and in-person workshop supports • Allows teachers to direct their own web-based learning using online resources, including videos of high-quality teaching • Was developed and tested with educators in state-funded preschools serving 4- and 5-year-olds	Undergoing efficacy trials to determine the impacts of MTP-M/S on teacher–student interactions and children's math and science knowledge and skills To learn more, visit http://www.mtpmathscience.net
Early Math Collaborative (previously known as Early Mathematics Education Project)	Jie-Qi Chen—Erikson Institute	• Targets comprehensive mathematics, with emphasis on increasing teachers' use of math language • Has 10 project-developed professional development modules • Professional development model includes workshops and in-person coaching • Was developed with pre-K and K teachers in the Chicago Public Schools	Currently being expanded to pre-K–3 Range of professional development services offered to schools and educational organizations To learn more, visit http://earlymath.erikson.edu
Supports for Science and Mathematics Learning in Prekindergarten Dual Language Learners (SciMath-DLL): Designing a Professional Development System	Kimberly Brenneman—Early Childhood STEM Lab at the National Institute for Early Education Research	• Targets both math and science • Pays particular attention to teachers of dual-language learners • Is designed to work with existing curricula • Professional development model includes workshops, in-person coaching, and school-based teacher workgroups • Is being developed and tested with educators in state-funded preschools serving 3-, 4-, and 5-year-olds in New Jersey	Entering final year of development and small-scale testing to determine effects on educators' attitudes and beliefs about science and math teaching and on their classroom practice in these domains To learn more, visit http://ecstemlab.org

(continued)

Table 7.1. (continued)

Project title	Lead researcher and institution	Selected key features	Development status
Professional Development for Culturally Relevant Teaching and Learning in Pre-K Mathematics	Anita Wager—University of Wisconsin–Madison	• Targets basic number operations • Uses "reciprocal funds of knowledge" approach in which cultural resources are leveraged in the classroom and families are helped to support math learning in the home • Teachers develop curriculum materials and teaching strategies as part of professional development • Provides 2 years of professional development that includes graduate-level coursework • Is being developed and tested with educators in Wisconsin's 4K (4-year-old kindergarten) program	Entering final year of development to assess the impacts on teachers' culturally relevant mathematics practice To learn more, visit http://www.wcer.wisc.edu/projects/projects.php?project_num=697

for PD to work. The most successful approaches tend to be those that involve higher dosages, are administered more intensively, and/or take place over a long period of time (e.g., Loeb, Rouse, & Shorris, 2007). The time frame here is many months, the full school year, or multiple years. For example, consider the reflective coaching cycle. It is intended to extend across school years, with mentor teachers or coaches working with teachers to provide models and resources for improved mathematics teaching and to facilitate reflection on practice in order to support improvement. To foster reflection, coaches and teachers meet to set goals, plan lessons, and engage in classroom observations (Costa & Garmston, 2002; Isner et al., 2011). Each teacher ideally takes time to think about the practices implemented and observed during a classroom lesson observation, and a postobservation reflection meeting occurs during which reflections are shared, progress toward goals is evaluated, and new goals are set. Then the cycle begins again. Similarly, workshop series intense enough to evoke change require many days, and a weeklong (or more) intensive seminar is not unusual. Coursework requires even more hours. The descriptions of TRIAD, MTP-M/S, and EMC illustrate the success that can occur when participants are engaged in ongoing PD across long periods of time. Still, an individual educator must be particularly dedicated (perhaps to pursue extra coursework), incentives must be high (PD credits, a degree, stipends), and/or administration must be supportive, flexible, and open to making the kinds of changes in school culture that allow protected time for teachers to reflect on their own practice and to learn with other adults.

Time in the Daily Schedule

On a day-to-day basis, time is a challenge, too. In my own work, educators consistently name time as the biggest barrier to participation in reflective coaching and teacher workgroups (Brenneman, Lange, & Stevenson-García, 2013). Recommendations that coaching reflection meetings occur in a quiet place for 60 minutes evoke laughter. In some situations, in-person coaching itself might not even be economically or practically viable due to the need for mentors to travel among sites. This concern applies to rural sites, but even in the city, two of my collaborating master teachers tell me they often feel like they spend more time finding parking than they do working with teachers in their community-based classrooms.

Similarly, the idea of finding 90 minutes for a teacher workgroup, when all can meet at the same time, draws rueful smiles, yet delving deeply into an issue of practice or a study of children's mathematical concept development in a workgroup or professional learning community (PLC) meeting takes a sizable block of dedicated time. If colleagues are together in one building, a 60-minute block might suffice, but if not, travel time to gather must be built into the scheduled time out of the classroom. If colleagues are to "work and learn together," they need regular and protected time to do this work and learning when all of them can leave their classrooms, knowing that they will not have to rush back to cover someone's lunch or lose their own lesson prep time. Sustained learning communities are likely to benefit both beginning and experienced teachers of early mathematics (see new National Science Foundation–funded work by Moss, Ball, and Boerst on the topic of developing elementary focused, collaborative assessment of practice at the NSF's web site), yet finding and shielding the time requires a concerted effort on the part of administrators to protect and advocate for this time (Annenberg Institute for School Reform, 2004).

One potential solution for time issues is the use of web-based PD resources. Although the specifics vary, all three successful ECME PD approaches described earlier in the chapter use technologically mediated supports for teachers. Using the web has some distinct advantages over face-to-face PD supports, which are time-consuming and must be squeezed into a busy day or occur outside of school hours at times that might not be convenient for the teacher. Online, self-guided learning, such as that used in MTP-M/S, allows teachers to individualize their instruction, spending time with the resources they believe will be most useful for their learning and development. Coaching supports also can be provided in a more flexible way when teaching practice is filmed and uploaded or sent to a coach who is not on-site (perhaps after initial meetings that occur in person to build rapport). Reflection and feedback meetings can occur via telephone, Skype, Google Hangout, or similar means. Similarly, Twitter and Facebook are increasingly being used for virtual communities of practice. A hybrid approach that blends in-person and web-mediated support has been found to be effective for PD focused on oral language development (Powell, Diamond, & Burchinal, 2012; see also Zaslow, Chapter 5), making it a good bet for ECME as well.

The ideal ratio of in-person and online contact and more fine-grained questions about the content and skills that are better developed through in-person or online methods are issues that researchers of PD in general are beginning to tackle. They will be no less important for ECME.

Other Curricular Needs

Although evidence is accruing about the importance of early mathematics for later achievement in math, literacy, and science (e.g., Duncan et al., 2007; Grissmer et al., 2010), there are other learning and teaching priorities in early childhood classrooms. Thus another aspect of time that affects math teaching is that there is "not enough time in the day." Practically, we must acknowledge that, although ECME PD is sorely needed, other content areas require attention too. Literacy and social-emotional goals often take precedence for teachers, and they tend to be more comfortable supporting these (Ginsburg, Lee, & Boyd, 2008; NRC, 2009), but even these areas in which teachers feel confident require support, as do other neglected areas such as science.

With limited time and resources, we have to balance needs and look for creative ways to help teachers find time to address math and other areas. MTP-M/S allows teachers to build skills in both math and science. The combination of BB/TRIAD with a curriculum that supports the development of self-regulation (Clements, Sarama, Layzer et al., 2012) or with research-based approaches to literacy, science, and social-emotional development (Clements, Sarama, Duke et al., 2012) might similarly enable teachers to develop skills in multiple domains. Helena Duch and colleagues at Columbia University are developing a curriculum and related PD that focus on language and literacy, math, and self-regulation, expanding to Head Start teachers an approach that has been successful in improving parents' math and literacy focused interactions with their children.[6]

Instead of adopting a new curriculum, one also might look to strengthening the mathematics portion of one's existing comprehensive curriculum (see the SciMath-DLL project description in later sections and the chapter appendix) or choosing a comprehensive curriculum that has a robust and proven mathematics component. As an example, the HighScope curriculum recently strengthened its mathematics offerings by introducing Numbers Plus (Epstein, 2009), and an upcoming efficacy study (funded by the Institute of Education Sciences; Larry Schweinhart, PI) will test the effects of the curriculum and related PD on teaching quality and on children's mathematics achievement. PD supports include 10 days of workshop training and 20 coaching visits focused on helping lead teachers understand math development and implement the Numbers Plus curriculum well.

The logic of these combined approaches is attractive, but more research is needed to fully explore the possible tradeoffs for math learning when other areas are introduced. For example, one would not want to lose the focused mathematics time that has been found to be important for learning (NRC, 2009), even as other curricular areas are addressed and combined with mathematics.

Personnel Issues Related to Long-Term, Sustained Professional Development

The types of professional support that "respond to an individual educator's unique background, experiences, teaching context, and role as an educator" (from earlier in this chapter) include mentoring and collegial workgroups. Coach–teacher or mentor–mentee partnerships are intended to be long-term to build trust between two individuals so that reflection occurs in an environment of respect, communication becomes more fluent, and knowledge builds over time. Likewise, professional learning communities have rules and protocols to support respectful interactions that facilitate communication and learning from one another. Time is required for rapport to be built and for groups to work productively together.

To the extent that these collegial relationships are disrupted, adult learning and development can suffer. Staffing changes are not unusual in school districts, and turnover in prekindergarten and early care settings is notoriously high, putting relationship-building at risk. If coaching/mentoring and teacher workgroups are critical to high-quality PD in ECME, then the relationships that grow among colleagues must be respected and protected. This does not mean that people should not be promoted or allowed to change positions, but it does suggest that administrators who are dedicated to improving early mathematics teaching and learning through individualized PD approaches must carefully consider personnel changes for their potential to disrupt the collegial relationships that are essential to the professional development of early care and education providers.

Individualization of Supports for Adult Learners

Hyson and Woods (Chapter 2) provide a comprehensive review of the varied demographic and educational backgrounds that characterize the early education work force. They also discuss the knowledge and beliefs that these adults hold regarding math generally and early childhood math specifically, detailing the ways that these beliefs and attitudes affect practice. Certainly these same characteristics will affect the ways that adults approach PD. Do they believe math is important for young learners, and are they motivated to seek more knowledge and skill in this area? Are they open to changing their practice? PD approaches need to be robust enough to work for motivated volunteers and for more reluctant learners if we are to effect large-scale change in teaching and learning. Doing so will require approaches that go beyond large group workshops to provide individualized supports.

Just as teachers are expected to assess the current knowledge and skills of their students as part of differentiating instruction, those of us who develop and deliver PD for adult learners must also develop better ways of assessing their current needs, knowledge, skills, and attitudes so that we can offer PD that is most effective for them. Formative and ongoing assessments that inform the goals and objectives of coaching, for example, and approaches that include learner-directed PD supports (such as MTP-M/S) provide ways to individualize PD offerings. Although the literature is not 100% clear that individual coaching is critical to PD success, Zaslow (Chapter 5) reports that, among studies that investigated links between coaching and classroom quality, about 80% provided positive evidence for the value of coaching. Similarly, Klein and Gomby (2008) note that the only study they reviewed that showed no effects of PD on instructional quality or child outcomes was the one approach that did not include a coaching component.

Early Childhood Mathematics Education Professional Development for All, Not Only Lead Teachers

We know that early childhood and early elementary school teachers need PD and support to improve their confidence and skills for teaching mathematics. Certainly, assistant teachers, who likely have lower education levels and less preservice and in-service math training than lead teachers, need supports as well. We need to engage and support the second adult in the room—if one even exists, which might not be the case in some care settings—but funding for PD and classroom staffing needs often make this difficult, if not out of the question. Creative approaches to staffing to allow for in-person training are needed, and web-based supports should be made available to teacher assistants. The latter, especially, could provide a relatively inexpensive way to support adult learners and, subsequently, child learners.

Educators outside of the classroom also require ECME PD. Ginsburg (2010) notes that university-based personnel who train preservice teachers, and perhaps provide in-service PD as well, often are not experts in early math learning (see also Schoenfeld & Stipek, 2011). School district math specialists vary in their knowledge of coaching strategies and mathematics content, which could influence their effectiveness in changing teacher attitudes, knowledge, and practice in ECME. This critical issue is the focus of a new National Science Foundation (NSF)–funded study that examines coaching in K–8 mathematics classrooms (Elizabeth Burroughs, PI). In preschool auspices, master teachers, coaches, and education supervisors who provide coaching supports for teachers are unlikely to be experts in early mathematics. PD for those who provide PD is also critical if they are to develop effective individualized supports for teachers in ECME.

Whereas some PD approaches provide project-specific coaches who are expert in ECME, others train other school personnel alongside teachers in early mathematics content and pedagogy with an eye toward building organizational capacity for sustainability. The Mount Holyoke Mathematics Leadership Program is an example of PD aimed at teachers (K and up), teacher leaders, and coaches that offers academic-year programs (site visits, seminars, and retreats), summer institutes, and online courses that build capacity within a district (Mount Holyoke College, n.d.). Having shown an impact on teachers' mathematical and pedagogical knowledge, the "Developing Mathematical Ideas" PD that is the basis for the Mount Holyoke Leadership Program is currently being evaluated for impacts on teachers' classroom practice and on student learning (funded by NSF; James Hammerman, PI). Among the ECME approaches highlighted elsewhere in this chapter, the SciMath-DLL approach (described in the sections that follow and in the chapter appendix) includes master teachers in training alongside current teachers, and the EMC provides informational sessions for administrators (although these are not described as PD).

Child Characteristics

As other authors in this volume have written, high-quality math teaching does not occur in many early childhood classrooms (see also NRC, 2009; Rudd et al., 2008), but this lack of excellence is more worrisome when the young children served are at risk for lack of school readiness due to factors such as low family socioeconomic status (SES) or a non-English home language. Unfortunately, low-SES children tend to receive instruction that is more focused on rote learning and skills than on cognitively stimulating math talk and reasoning (Stipek & Byler, 1997). On the positive side, much of the current research in early math learning involves at-risk learners, and research of PD programs includes those who teach these learners, so we know that well-researched programs to improve mathematics teaching and children's math learning are available.

In contrast, issues of language and culture have been less frequently addressed as they relate to mathematics learning and teaching in the early years. This is true for language and literacy PD studies as well (Zaslow et al., 2010b). As Zaslow points out (Chapter 5), a PD approach can "flounder if it does not take into account key characteristics of the population in a classroom or program, such as the concentration of dual- language learners."Demographic studies tell us that the population of students with immigrant parents will increase by 60% over the next 15 years (Passel, 2007). Our nation's young students increasingly speak a language other than English at home, with almost half of all English language learners concentrated in pre-K to Grade 3 (Matthews & Ewen, 2005). The achievement gap between children with a non-English home language and those with English as their first language, measured by the National Assessment of Educational Progress, is substantial. Research suggests that Hispanic children enter kindergarten well behind non-Hispanic peers in literacy and math (Denton & West, 2002), that gaps persist into the upper grades (Perie, Grigg, & Dion, 2006), and that the disparity is even greater for Hispanic dual-language learners and English language learners compared to their English-proficient Hispanic peers (Laosa & Ainsworth, 2007).

Although a non-English home language is a risk factor for lack of school readiness and achievement, students learning multiple languages also possess multiple resources to draw on, rather than just obstacles to overcome, as they participate in the classroom community. Interesting math learning experiences can stimulate young children to communicate with their peers and their teachers in both English and the home language. To illustrate, I recently observed a measurement lesson as part of a PD research project I am conducting. As the teacher asked children questions about her length measuring procedure (with some mistakes thrown in as a pedagogical strategy to stimulate conversation), one child stated her opinion with just a single word. That does not seem like much conversation, perhaps, but she had been so silent to that point in the year, that her word prompted the child next to her to say, with no small amount of wonder, "She talked!" Although any interesting lesson might have prompted this first verbal step, it was a math lesson that did. ECME PD that recognizes and supports teachers to use mathematics as a powerful tool for developing language and conversation skills could develop readiness beyond math; when children are encouraged to communicate—and given something they want to communicate about—mathematics has the potential to support growth in both languages as well as in mathematics reasoning and skills. Recall that the TRIAD work found clear benefits of high-quality math teaching on language learning. The researchers hypothesized that higher levels of teacher math talk were part of the reason that children's language skills improved.

As with math, early education teachers are likely to need PD supports for dual-language learning because they are ill equipped to meet children's specific learning needs, largely due to lack of appropriate coursework in teacher preparation programs (Freedson, 2010). This challenge must be acknowledged and addressed. Put simply, we need to design, choose, and use ECME PD that prepares educators to teach the children in their classrooms now as well the ones who will fill their classrooms and centers in the future. These children are increasingly diverse in their linguistic and cultural backgrounds, and ECME approaches that acknowledge this will be required.

PROMISING APPROACHES TO EARLY CHILDHOOD MATHEMATICS EDUCATION FOR DIVERSE LEARNERS IN DIVERSE LEARNING SETTINGS

Two projects for diverse learners are highlighted here. Although each is in the relatively early stages of design, implementation, and testing of their particular approaches to ECME, their explicit aim is to support linguistically and culturally diverse learners and their teachers.

Supports for Science and Mathematics Learning in Prekindergarten Dual-Language Learners: Designing a Professional Development System[7]

Features of the Approach

Similar to MTP-M/S, SciMath-DLL addresses both mathematics and science, aiming to leverage connections between the two domains (e.g., in measurement and pattern detection) at the same time that language and literacy are supported. As part of SciMath-DLL PD, preschool educators in state-funded preschool programs serving 3-, 4-, and 5-year-olds are provided with concrete examples of the ways that science and mathematics learning experiences support language and literacy learning. For instance, these interactions support vocabulary development by exposing children to new words in meaningful contexts as children describe what they are doing (observe, predict, estimate, sort, experiment), what they are observing (chrysalis, roots, seed pods, parallelogram), and the attributes of objects and events (sticky, dirty, roundish, pointy, more than, and less than). It is expected that teachers will be more likely to devote time to math and science learning experiences when they see that these can also support language and literacy (Buxton, Lee, & Santau, 2008).[8]

The SciMath-DLL approach was codeveloped with educators using three different curriculum models and does not require the adoption of a new, math-specific curriculum. Acknowledging the effectiveness of linking teacher PD closely to curriculum for math, science, and ELL (Buxton et al., 2008; Lee, 2005; NRC, 2009), SciMath-DLL PD provides multiple opportunities for educators to combine the general principles of the SciMath-DLL approach with their school's curriculum, to reflect on the ways in which this is easy or difficult, and to document these experiences as part of the ongoing cycle of evaluation and improvement.

Teachers are provided with workshop experiences that support the development of knowledge about children's learning in these areas and that provide practical strategies for improving instructional offerings through connected learning experiences that center on a particular math concept (e.g., shape recognition or subitizing). These active learning experiences engage teachers with model lesson plans that detail the specific decisions the developers made as they created that plan. Sets of plans serve both as models for teachers to create their own sets of learning experiences that support particular math (and science) concepts and as "something to do on Monday" with their students. In addition to 3–4 workshops throughout the school year, the SciMath-DLL model includes individualized, classroom-based coaching using district master teachers and cocoaching with project experts in early math and science teaching and learning. Teacher workgroups, in which 6–8 colleagues and their master teacher discuss a math teaching challenge or explore children's learning of math concepts, are convened approximately every 6 weeks.[9]

Research Results

Now in its fourth year, SciMath-DLL has successfully engaged teachers and master teachers as collaborators in developing content, procedures, and supporting materials for workshops, model lesson plans, teacher workgroups, and individualized practice-based coaching. Major implementation challenges included the novelty of the coaching cycle, the amount of paperwork, and time limitations. Educators noted that even with challenges, the SciMath-DLL approach was valuable because it sanctioned and "forced" reflection on practice, which they value. Teachers and coaches noted improved attitudes toward math and science teaching, increased appreciation for reflection, increased awareness that lessons are often overstuffed, and less directive teaching (Brenneman et al., 2013). Coaches reported learning science, technology, engineering, and math (STEM) content, better recognition of effective teaching methods, and improved understanding of the reflective coaching cycle, results that suggest the importance of attending to the PD of PD providers. If the SciMath-DLL approach shows positive effects on instructional quality and teacher self-report of change, effects on child outcomes will be measured in subsequent work.

Professional Development for Culturally Relevant Teaching and Learning in Pre-K Mathematics[10]

Features of the Approach

This research and development project creates and studies a PD model and resources that focus on culturally relevant teaching and learning. Basic number operations are stressed, with the intent of helping teachers explore this area in depth. Using PLCs and coursework, the project employs a reciprocal "funds of knowledge" framework through which children's out-of-school experiences are accessed and leveraged to provide instruction that is culturally relevant (Moll, Amanti, Neff, & González, 1990, 2005) at the same time that teachers are helped to support families to better foster mathematical learning outside of school.

The PD approach builds on the research team's prior work with Cognitively Guided Instruction (CGI; Carpenter, Fennema, & Franke, 1996; Fennema et al., 1996; see also new Institute of Education Sciences–funded work that studies the ways that principal, teacher, and student characteristics interact with a CGI-based approach to PD for Grades 1 and 2 math to affect student achievement,

Robert Schoen, PI). In the CGI model, no instructional materials or specific teaching strategies are promoted; instead, teachers are guided to develop their own (or adapt other materials) based on observing and interacting with students and reflecting on what they have observed.

The innovation of the University of Wisconsin–Madison work lies in its focus on helping teachers understand how children's mathematical strategies and knowledge are influenced by both school and out-of-school experiences, how mismatch between these can cause difficulty, and how teachers can bring children's experiences into better alignment across school and home settings. The underlying idea is that some portion of the achievement gap can be accounted for by this mismatch. Reducing the mismatch will allow children to become more mathematically proficient and will lessen the achievement gap.

Cohorts of teachers are engaged in 2 years of PD that include 4 graduate-level courses. Early in the project, participants learn about mathematical development and then apply what they learn to practice. Every other week, teachers engage in seminars that focus on early number operations, developmentally appropriate instructional practices, and the home–school connection. During the second year, teachers engage in an action research project in which they gather monthly to address an element of classroom action research related to the research questions of the PD research project (e.g., How can a PD program help establish and support a teacher–family learning community for culturally and developmentally responsive mathematics?). As new cohorts are added to the project, experienced teachers can become mentors to new ones.

Research Results

The PD development team has documented the wealth of knowledge and resources available in children's homes, allowing teachers to better plan instruction that maps onto these experiences should they wish (A. Wager, personal communication, November 3, 2012). Preliminary findings show that participating teachers broaden their ideas about children's understanding of mathematics and about home resources that foster these understandings. Wager (in press) reports on the experiences of one teacher involved in the project and the wide-ranging effects on her practice as she seeded her classroom with mathematical tools and manipulatives, incorporated mathematics into daily routines and transitions, provided planned mathematical learning experiences for children, and responded spontaneously to children's mathematical play and interests. As the year progressed, children grew in their numeracy skills, problem-solving skills, and use of math language during routines.

A WISH LIST TO INFORM EARLY CHILDHOOD MATHEMATICS EDUCATION PROFESSIONAL DEVELOPMENT RESEARCH, POLICY, AND PRACTICE

With the recognition that young math learners are an increasingly diverse group, culturally and linguistically, comes the need to design careful, empirical studies that focus on children learning two languages and the specific ways that the cognitive and social challenges (and benefits) of being a dual-language learner affect and are affected by mathematics learning and development. This is just one of many wishes that I, and others, have for ECME PD research, policy, and practice.

Compare Effectiveness in Rigorous Ways

Experimental study of various forms and schedules of ECME PD will allow us to begin to tease apart the particular aspects of a PD approach that make it effective under particular circumstances (Klein & Gomby, 2008). Many PD interventions are compared with business-as-usual practices (which is nothing in many pre-K classrooms when it comes to math); thus results do not tell us much more than that a given approach is better than nothing. At the same time, an approach that incorporates multiple components (e.g., workshops, coaching, and professional learning communities) might be found to be effective, but it is impossible to determine whether one or more components are the active ones that contribute to change and how the components might interact to bring about change. (For an in-depth treatment of this issue, see Chapter 8.) Zaslow and colleagues (2010a, 2010b) note the general lack of rigor in PD studies. More carefully designed studies of PD, and of ECME PD specifically, could elucidate which delivery models are most effective with various types of educator-learners, based on characteristics such as years in the profession, level of education, mathematical knowledge, and so forth.

Benefit–Cost Analyses

A related issue concerns the costs associated with different forms and components of PD. "Canned" workshops are less expensive and time-consuming than college coursework or ongoing, in-person coaching supports; however, personalized, classroom-based, and/or sustained approaches to PD are more effective than one-shot workshops, or even a series of these, for changing practice and improving instructional quality (Klein & Gomby, 2008; Loeb et al., 2007). In early childhood education, the argument has been made—convincingly—that investments in preschool education payoff in the long term because children who get that better start are more likely to graduate from high school, get higher-paying jobs, pay more taxes, stay out of prison, among other positive outcomes (e.g., Barnett & Masse, 2007; Temple & Reynolds, 2007). An economic analysis of the benefits for higher instructional quality and learning outcomes for students versus costs of various forms of PD would be of great use to policy makers. Cheap PD is not cheap if no benefit accrues in children's learning. On the other hand, intuitively, it would seem that more intense coaching (say, weekly versus monthly) would yield greater benefit for child learning, but this more expensive level of intensity might not yield increased benefits (see Ramey & Ramey, 2006). Not surprisingly, given the lack of studies that directly compare different PD strategies, no benefit–cost analysis exists for early childhood math PD specifically or for PD in general (W.S. Barnett, personal communication, November 18, 2012).

Greater than the Sum of the Parts Or . . . ?

As discussed earlier, combining mathematics PD with PD for other critical areas of child learning and development might take advantage of connections among domains to boost adult and child learning in multiple areas. The question remains whether approaches such as this work in a positive way to foster multiple areas of growth and development, or whether mathematics teaching and learning is better supported when it is the sole focus of PD. Some marriages might be more successful than others, so different (theoretically motivated) combinations should be explored. Research that directly compares multidomain approaches with single-domain approaches would be of practical interest. Can we support multiple domains well without losing the benefits of a focus on math? If so, we might be able to get more bang for our limited PD buck.

Bridge the Divide Between Pre-K and Elementary School

My search of currently funded projects in the research pipeline was not exhaustive; however, it is clear that few projects cross the boundaries between prekindergarten and early elementary grades. Articulation across grades is understood to be desirable in K–12 education, yet this approach has not extended from K–12 to prekindergarten for the most part (Schoenfeld & Stipek, 2011). Many school districts provide prekindergarten education; thus the potential exists for articulation between pre-K and kindergarten. Such approaches could offset some of the "fadeout" that might occur as children who have received high-quality preschool experiences move into early elementary grades that might not be of similar quality (Barnett, 2004). Involving teachers across grades in the same types of effective PD experiences (as in the expanded EMC project, the follow-through condition, or TRIAD, described earlier), has the potential to reduce this loss and perhaps even allow children's learning gains to increase over time. The adoption of the Common Core and resulting efforts by some states to align their early learning standards to the kindergarten and early elementary standards could also support increased alignment.

Engage End Users from the Start and Throughout Development

Designers of ECME PD should engage educators from the beginning of the development process. Although it takes time to build these collaborative relationships, and it can be tricky to negotiate a shared language and shared goals, PD research and development partnerships that engage end users from the beginning are more likely to result in materials that teachers will use because they will be practical, responsive to their needs, and more closely aligned with the realities of life in the teaching trenches.[11] Soliciting and being responsive to feedback from educators is a way for PD designers and providers to improve their own practice. In my own work, I find myself saying, "We are not here to do this *to* you or to *give* you PD. We are here to offer something and to get your critical and honest feedback about whether it helps you. We need you, and we are doing this with you." To create the most collegial and productive environments for learning, PD developers and providers should be vigilant neither to switch into one-way expert provider mode nor to allow teachers to fall back on their usual experiences and expectations for PD in which we "tell them how to do it."

CONCLUSION

Although my wish list reflects the research base and others' recommendations (e.g., Klein & Gomby, 2008; NRC, 2009; Zaslow, Chapter 5), it is also informed and colored by personal experience in schools collaborating on educational research after starting life as a developmental psychologist specializing in cognitive development. Reading and reflecting on the literature, my experiences with teachers, and discussions with colleagues, I cannot help but be a bit overwhelmed by the task before us. To quote Herbert P. Ginsburg (2010), "This is a massive job, but unless we do it, we will not succeed in implementing effective early math education." And it is clear that we must do this job. When we fail to prepare and support their teachers, we fail children. Effective early math education is vital for reasons of equity, economic viability, and human capital development. For global economic competitiveness, we must foster a STEM-literate citizenry and support our increasingly diverse student body to become the mathematicians, engineers, technology innovators, physicians, and scientists of the future. If we, as providers, researchers, and consumers of ECME PD, do our jobs well now, we can ensure that today's 4- and 5-year-olds can become not just STEM literate but—if they wish—preschool and elementary teachers who are mathematically competent and confident, who see themselves as capable math learners, and who are ready, willing, and able to prepare the next generation of STEM learners.

FOR REFLECTION AND ACTION

1. Brenneman summarizes guidelines for in-service professional development in mathematics taken from the NRC's 2009 report. Reading through those guidelines, how would you rate your professional development programs (either those you plan and deliver or those you have experienced)? Where are the strengths, and where are the needs for improvement?

2. Brenneman summarizes the characteristics of three well-evaluated approaches to in-service professional development in ECME. Depending on your interests and needs, you might want to learn more about one or more of these using the chapter references and web links. Are there specific features of some of these approaches that you might wish to adopt?

3. An interesting aspect of several of the PD approaches summarized by Brenneman is that they have combined mathematics-related PD with other domains such as science, language and literacy, self-regulation, and enhanced teacher–child interactions. What do you see as the advantages, or possibly the challenges, of this kind of inclusive PD?

4. Brenneman outlines some commonly encountered obstacles to successfully implementing in-service professional development in mathematics. How do these map

onto your experience? What approaches may have been successful in removing these obstacles?

5. Toward the end of the chapter, Brenneman addresses the need to build teachers' capacity to provide high-quality math education to dual-language learners and children in immigrant families. If this is a group of learners upon whom your work focuses (or will increasingly focus in the future), you might want to look further into the PD examples Brenneman describes, perhaps with a group of colleagues in a professional learning community.

CHAPTER 7 NOTES

1. Table 7.1 provides a quick summary of the key features of these three approaches as well as two newer PD models that are still under development.
2. University at Buffalo, SUNY and University of Denver, Julie Sarama and Douglas Clements (Principal Investigators [PIs]).
3. Center for Advanced Study of Teaching and Learning, University of Virginia, Mabel Kinzie (PI).
4. Erikson Institute, Jie-Qi Chen (PI).
5. "In the pipeline" approaches to ECME PD described here were identified by searching the National Science Foundation (NSF) and Institute for Education Sciences (IES) web sites. Each site's search functions were used to locate research projects that potentially involved pre-K to Grade 1 in-service math professional development for educators. Search terms such as "math," "professional development," "in-service," and "early education" were used to identify potential candidates within relevant programs at each agency (e.g., the Discovery Research K–12 program at NSF and the Effective Teachers and Effective Teaching, Mathematics and Science Education program at IES). Abstracts and titles were used to further narrow the field among projects funded in the last 5 years. When grade levels could not easily be determined from the funding agency web sites, a search for a project web site was done to clarify.
6. More information about Duch and colleagues' work can be found on the NSF's site at http://www.nsf.gov.
7. National Institute for Early Education Research, Rutgers University, Kimberly Brenneman (PI).
8. Because there is evidence that focused math time—that which does not involve other content areas—is important for math learning (NRC, 2009), teachers also are supported to provide focused math experiences for learners.
9. Please see the Chapter 7 appendix for an example of a SciMath-DLL model lesson plan and supporting materials.
10. University of Wisconsin–Madison, Anita Wager (PI).
11. Barnes and colleagues (2010) is a useful resource for developing and nurturing partnerships among researchers and educators written by those who have engaged in such collaborations around math and science education.

REFERENCES

Annenberg Institute for School Reform. (2004). *Professional learning communities: Professional development strategies that improve instruction.* Providence, RI: Brown University. Retrieved from http://annenberginstitute.org/sites/default/files/product/270/files/ProfLearning.pdf

Barnes, D., Benenson, G., Heuer, L., Hobbs, M., King, K., Kinzer, C. . . . Wiberg, K. (2010). *Fostering knowledge use in STEM education: A brief on R&D partnerships with districts and schools.* Washington, DC: Community for Advancing Discovery Research in Education.

Barnett, W.S. (2004). Does Head Start have lasting cognitive effects? The myth of fade-out. In E. Zigler & S. Styfco (Eds.), *The Head Start debates* (pp. 221–249). Baltimore, MD: Paul H. Brookes Publishing Co.

Barnett, W.S., & Masse, L.N. (2007). Early childhood program design and economic returns: Comparative benefit-cost analysis of the Abecedarian program and policy implications. *Economics of Education Review, 26*(1), 113–125.

Birman, B.F., Desimone, L., Porter, A.C., & Garet, M.S. (2000). Designing professional development that works. *Educational Leadership, 57*(8), 28–33.

Brenneman, K., Lange, A.A., & Stevenson-García, J. (2013, April). *Improving math and science supports for at-risk preschool children by supporting the teachers who teach them.* Paper presented at the biennial meeting of the Society for Research in Child Development, Seattle, WA.

Buxton, C., Lee, O., & Santau, A. (2008). Promoting science among English language learners: Professional development for today's culturally and linguistically diverse classrooms. *Journal of Science Teacher Education, 19*, 495–511.

Carpenter, T.P., Fennema, E., & Franke, M.L. (1996). Cognitively Guided Instruction: A knowledge base for reform in mathematics instruction. *Elementary School Journal, 97*(1), 3–20.

Center for Advanced Study of Teaching and Learning. (n.d.). *MyTeachingPartner—Mathematics/Science.* Retrieved from http://curry.virginia.edu/research/centers/castl/project/mtp-math-science

Clements, D.H., & Sarama, J. (2008). Experimental evaluation of the effects of a research-based preschool mathematics curriculum. *American Educational Research Journal, 45*(2), 443–494.

Clements, D., Sarama, J., Duke, N.K., Hemmeter, M.L., & Brenneman, K. (2012, April). *Connect4Learning: Early education in the context of mathematics, science, literacy, and social emotional development.* Paper presented at the annual conference of the American Educational Research Association, Vancouver, British Columbia, Canada.

Clements, D.H., Sarama, J., Layzer, C., Unlu, F., Bedrova, E., & Leong, D. (2012, March). *Efficacy of an intervention synthesizing scaffolding designed to promote self-regulation with and early math curriculum: Effects on executive function.* Paper presented at the Society for Research in Educational Effectiveness, Washington, DC.

Costa, A.L., & Garmston, R.J. (2002). *Cognitive coaching: A foundation for renaissance schools.* Norwood, MA: Christopher-Gordon.

Denton, K., & West, J. (2002). *Children's reading and mathematics achievement in kindergarten and first grade* (NCES 2002–125). Washington, DC: National Center for Education Statistics.

Duncan, G.J., Dowsett, C.J., Claessens, A., Magnuson, K., Huston, A.C., Klebanov, P., . . . Japel, C. (2007). School readiness and later achievement. *Developmental Psychology, 43*(6), 1428–1446.

Epstein, A.S. (2014). *The intentional teacher: Choosing the best strategies for young children's learning.* Rev. ed. Washington DC: National Association for the Education of Young Children.

Epstein, A.S. (2009). *Numbers Plus preschool mathematics curriculum.* Ypsilanti, MI: HighScope Press.

Fennema, E., Carpenter, T.P., Franke, M.L., Levi, L., Jacobs, V.R., & Empson, S.B. (1996). A longitudinal study of learning to use children's thinking in mathematics instruction. *Journal for Research in Mathematics Education, 27*(4), 403–434.

Freedson, M. (2010). Educating preschool teachers to support English language learners. In E.E. Garcia & E. Frede (Eds.), *Young English language learners: Current research and emerging directions for practice and policy* (pp. 165–183). New York, NY: Teachers College Press.

Ginsburg, H.P. (2010, February). *Professional development for early childhood mathematics education.* Paper presented to the STEM Summit, Irvine, CA. Retrieved from http://www.slideshare.net/stemsummit/professional-development-for-early-childhood-mathematics-education-herbert-ginsburg

Ginsburg, H.P., Lee, J.S., & Boyd, J. (2008). Mathematics education for young children: What it is and how to promote it. *Society for Research in Child Development Social Policy Report, 22*(1), 1–23.

Grissmer, D., Grimm, K.J., Aiyer, S.M., Murrah, W.M., & Steele, J.S. (2010). Fine motor skills and early comprehension of the world: Two new school readiness indicators. *Developmental Psychology, 46*(5), 1008–1017.

Isner, T., Tout, K., Zaslow, M., Soli, M., Quinn, K., Rothenberg, L., & Burkhauser, M. (2011). *Coaching in early care and education programs and Quality Rating and Improvement Systems (QRIS): Identifying promising features.* Report prepared for the Children's Services Council of Palm Beach County. Washington, DC: Child Trends.

Klein, A., & Starkey, P. (2004). Fostering preschool children's mathematical knowledge: Findings from the Berkeley Math Readiness Project. In D.H. Clements, J. Sarama, & A.-M. DiBiase (Eds.), *Engaging young children in mathematics: Standards for early childhood mathematics education.* Mahwah, NJ: Lawrence Erlbaum Associates.

Klein, L.G., & Gomby, D.S. (2008, October). *A synthesis of federally-funded studies on school readiness: What are we learning about professional development.* Working paper prepared for A Working Meeting on Recent School Readiness Research: Guiding the Synthesis of Early Childhood Research, Washington, DC.

Klibanoff, R., Levine, S.C., Huttenlocher, J., Vasileyva, M., & Hedges, L.V. (2006). Preschool children's mathematical knowledge: The effect of teacher "math talk." *Developmental Psychology, 42*(1), 59–69.

Laosa, L.M., & Ainsworth, P. (2007, April). *Is public pre-K preparing Hispanic children to succeed in school* (Preschool Policy Brief, Issue No. 13). New Brunswick, NJ: National Institute for Early Education Research.

Lee, O. (2005). Science education with English language learners: Synthesis and research agenda. *Review of Educational Research, 75*(4), 491–530.

Lee, Y., Kinzie, M.B., & Whittaker, J. (2012). Impact of on-line support for teachers' open-ended questioning in pre-K science activities. *Teaching and Teacher Education, 28*(4), 568–577.

Loeb, S., Rouse, C., & Shorris, A. (2007). Introducing the issue: Excellence in the classroom. *The Future of Children, 17*(1), 3–14.

Matthews, H., & Ewen, D. (2005). *Reaching all children? Understanding early care and education participation among immigrant families.* Washington, DC: Center for Law and Social Policy.

McCray, J.S., & Chen, J-Q. (2011). Foundational mathematics: A neglected opportunity. In B. Atweh, M. Graven, W. Secada, & P. Valero (Eds.), *Mapping equity and quality in mathematics education* (pp. 253–268). Dordrecht, Netherlands: Springer.

Moll, L.C., Amanti, C., Neff, D., & González, N. (1992). Funds of knowledge for teaching: Using a qualitative approach to connect homes and classrooms. *Theory into Practice, 31*(2), 132–141.

Moll, L., Amanti, C., Neff, D., & González, N. (2005). Funds of knowledge for teaching: Using a qualitative approach to connect homes and classrooms. In N. González, L.C. Moll, & C. Amanti (Eds.), *Funds of knowledge: Theorizing practices in households, communities, and classrooms* (pp. 71–87). Mahwah, NJ: Lawrence Erlbaum Associates.

Mount Holyoke College. (n.d.). *Mathematics leadership programs.* Retrieved from https://www.mtholyoke.edu/cpd/summermathteachers

National Research Council. (2009). *Mathematics learning in early childhood: Paths toward excellence and equity.* Committee on Early Childhood Mathematics, C.T. Cross, T.A. Woods, & H. Schweingruber (Eds.). Center for Education, Division of Behavioral and Social Sciences and Education. Washington, DC: National Academies Press.

Passel, J. (2007). *Unauthorized migrants in the United States: Estimates, methods, and characteristics.* Retrieved from https://www1.oecd.org/els/39264671.pdf

Perie, M., Grigg, W., & Dion, G. (2006). *The nation's report card: Mathematics 2005* (NCES 2006–453). Washington, DC: National Center for Education Statistics.

Pianta, R.C., La Paro, K.M., & Hamre, B. (2006). *Classroom Assessment Scoring System (CLASS).* Charlottesville, VA: Center for Advanced Study of Teaching and Learning.

Powell, D.R., Diamond, K.E., & Burchinal, M. (2012). Using coaching-based professional development to improve Head Start teachers' support of children's oral language skills. In C. Howes, B.K. Hamre, & R.C. Pianta (Eds.), *Effective early childhood professional development: Improving teacher practice and child outcomes* (pp. 13–29). Baltimore, MD: Paul H. Brookes Publishing Co.

Ramey, S.L., & Ramey, C.T. (2006). How to create and sustain a high quality work force in child care, early intervention, and school readiness programs. In M. Zaslow

& I. Martinez-Beck (Eds.), *Critical issues in early childhood professional development* (pp. 325–368). Baltimore, MD: Paul H. Brookes Publishing Co.

Rudd, L.C., Lambert, M.C., Satterwhite, M., & Zaier, A. (2008). Mathematical language in early childhood settings: What really counts? *Early Childhood Education Journal, 36,* 75–80.

Sarama, J., Clements, D.H., Starkey, P., Klein, A., & Wakeley, A. (2008). Scaling up the implementation of a pre-kindergarten mathematics curriculum: Teaching for understanding with trajectories and technologies. *Journal of Research on Educational Effectiveness, 1*(2), 89–119.

Sarama, J., Lange, A.A., Clements, D.H., & Wolfe, C.B. (2012). The impacts of an early mathematics curriculum on oral language and literacy. *Early Childhood Research Quarterly, 27*(3), 489–502.

Schoenfeld, A.H., & Stipek, D. (2011). *Math matters: Children's mathematical journeys start early.* Retrieved from http://earlymath.org/earlymath/wp-content/uploads/2012/03/MathMattersReport.pdf

Stipek, D.J., & Byler, P. (1997). Early childhood teachers: Do they practice what they preach? *Early Childhood Research Quarterly, 12*(3), 305–325.

Temple, J.A., & Reynolds, A.J. (2007). Benefits and costs of investments in preschool education: Evidence from the Child-Parent Centers and related programs. *Economics of Education Review, 26*(1), 126–144.

Wager, A.A. (2013). Practices that support mathematics learning in a play-based classroom. In L. English & J. Mulligan (Eds.), *Reconceptualizing early mathematics learning.* New York, NY: Springer.

Woodcock, R.W., McGrew, K.S., & Mather, N. (2001). *Woodcock-Johnson III.* Rolling Meadows, IL: Riverside Publishing.

Zaslow, M., Tout, K., Halle, T., Vick Whittaker, J.E., & Lavelle, B. (2010a). Emerging research on early childhood professional development. In S.B. Neuman & M.L. Kamil (Eds.), *Preparing teachers for the early childhood classroom: Proven models and key principles* (pp. 19–47). Baltimore, MD: Paul H. Brookes Publishing Co.

Zaslow, M., Tout, K., Halle, T., Whittaker, J.V., & Lavelle, B. (2010b). *Towards the identification of features of effective professional development for early childhood educators.* Prepared for Policy and Program Studies Service, Office of Planning, Evaluation and Policy Development, U.S. Department of Education. Retrieved from http://www.ed.gov/about/offices/list/opepd/ppss/reports.html

ACKNOWLEDGMENTS

The writing of this chapter was supported, in part, by grant number DRL-1019576 from the National Science Foundation. The opinions, findings, and conclusions or recommendations expressed in this material are those of the author and do not necessarily reflect the views of the National Science Foundation. I thank my colleagues Alissa Lange and Jorie Quinn in the Early Childhood STEM Lab at the National Institute for Early Education Research for their collaborative spirit. Ellen Frede, Judi Stevenson-García, and Alex Figueras-Daniel helped to envision SciMath-DLL, and it is better for their having been involved. I am fortunate to work with many dedicated and talented educators on the SciMath-DLL project. I appreciate their partnership and extend special thanks to Jennifer Rubin, Lily Rajan, and Kadejah Davis for sharing the classroom stories described in the chapter appendix. I offer my deep gratitude to Marilou Hyson, Taniesha A. Woods, and Herbert P. Ginsburg for their patience and thoughtful comments on outlines and drafts.

7

Appendix

Science and Mathematics Learning in Prekindergarten Dual-Language Learners

Our goal when we design and implement SciMath-DLL professional development workshops[1] is to provide preschool teachers with some basic information about children's learning in a key area of mathematics (or science) through slides and discussion but to spend the bulk of our time engaging in small-group learning activities with the following characteristics:

- Centers on a draft lesson plan that can guide our adult learning experience and that can be used with children

- Is active and engaging for adult learners

- Models intentional lesson planning by including written annotations that let teachers know why we designed the activity as we did and supporting discussion that encourages them to ask questions, and allows them to suggest changes or extensions to the lesson

- Supports thinking about differentiation for learners with different needs (especially children who are at different levels of first- and second-language development)

- Illustrates important math (and science) concepts and skills and supports reflection on children's learning of these concepts and skills

- Connects to other small-group lessons plans—which are also provided, demonstrated, and discussed in the workshop—to illustrate the importance

---

1 From National Institute for Early Education Research. (2013). *SciMath-DLL: Introduction, Sample Lesson Plan, and Classroom Stories.* New Brunswick, NJ: Author; reprinted by permission.

of engaging children in multiple, conceptually linked learning experiences to promote deeper learning

The materials provided here include a sample lesson plan and two "reports from the field" after classroom implementation. The lesson plan describes an activity, Roll and Build, that was part of a module about number and operations. The plan includes our notes on potential variations, some of our reasons for creating the lesson as we did, and discussion questions for the workshop. A number of teachers reported trying some version of this lesson with their students. Two teachers shared their experiences. One wrote up her notes, which are summarized here. Another provided a video of her lesson, parts of which are described and transcribed here. Teachers' feedback during workshop discussions and after they try lessons in their classrooms are used to edit the lesson plans to make them more practical and informative for the next group of educators who will use them. The plan provided here, for example, will likely change based on teachers' experiences and other expert feedback through an iterative review process. Still, this version provides an illustration of our approach. For more information, please visit http://ecstemlab.org.

SAMPLE LESSON PLAN

Small Group: Roll and Build (Includes Writing Numerals)

Description

Children work in a small group to create a structure out of blocks. They take turns rolling a die, identifying the number, writing down the numeral represented on the die, and adding that number of blocks to the structure.

Learning Objective(s)

- To count accurately in the small number range
- To identify numbers in various representational formats
- To represent numbers in various ways

Materials

- One die
- Blocks for building (foam, unit blocks, or tabletop blocks)
- Paper and pencil

Procedure

1. Explain the procedure for the game. Each child will get a turn to do different jobs. The roller rolls the die and identifies the number represented. The builder adds that number of blocks to the structure. The writer writes down the number of blocks added to the structure.

2. Choose one child to roll first. Choose another to build and a third to write.

3. The first child rolls the die and identifies the number that is represented. (Other children can help as needed.) The builder begins a block structure using the same number of blocks as was shown on the die. At the same time, the writer represents the number on paper with a written numeral.

4. On the next turn, new children take over the jobs. As this continues, the structure will get larger and larger. (Note: If switching every job is difficult for children, allow one child to roll and build while all others help with counting or writing the numerals. You might also have one child roll while others count blocks and build their own structures.)

5. When it is just about time to finish, ask children to think about how many blocks are in their structure. Can they estimate?

6. Ask how they could find out for sure. If children do not suggest it, use the written numerals as one way to find out how many blocks were used. (The teacher will have to do the adding here, but the more important concept is that the written numbers allow the children to know how many blocks are in the structure without taking it apart.)

7. When you are all done, you might work together to count the blocks as you take apart the structure.

Discussion Questions During Workshop Activity

- How could you adjust this game for children who need more support in various aspects of numerical development (e.g., counting, numerical operations such as addition, writing numerals)?

- Writing numerals is not necessary for the basic Roll and Build activity, but it provides a way to practice writing numerals for a purpose. Children might have difficulty with this aspect of the lesson, however. What are some other ways they might keep track of numbers without writing actual numerals?

CLASSROOM STORIES

The first classroom story comes from a teacher of 3- and 4-year-olds in a state-funded preschool program in New Jersey. The teacher used the basic Roll and Build lesson more than once, reflecting on its implementation each time and adapting it to meet the needs of her children and the learning objectives she had for them.

> I have used the activity Roll and Build in my classroom a few times this year with my three year old students and it was very successful. I modified the lesson in order to make it appropriate for my specific students. I have done this activity as a whole group and also small group. We used one die and each child got a turn to count the blocks and add the correct number of blocks.
>
> The first time we played the game in a whole group each child rolled the die and counted the dots on the die and then added the same number of blocks to the structure. After each student had a turn (15 students) we observed the structure we built. I asked the students to discuss and share what they thought the structure looked like. The next time we played the game I introduced the role of the jobs: roller, builder, writer.

I felt that for my particular group of students it was difficult for them to follow the jobs especially in a whole group. So the next time I played the game I chose to do it in a small group with four students. I couldn't believe the difference between doing the same activity in a small group versus whole group. The four students were engaged and excited about what we were creating. We discussed how many blocks we were using and they estimated how many we used altogether. I felt it was a perfect activity to do in a small group with my students.

I also tried it again in a whole group and this time we decided we were going to choose something we wanted to build and our goal was to build it together. My students chose to build a train on train tracks. I was so impressed with their strategies for placing the blocks and how it really looked like a train on tracks. Overall I feel that it is an excellent activity to use with young children because they are excited to create and build with blocks but the activity involves the students counting, estimating, and so forth. The students enjoyed playing the game and were learning at the same time, it was great!

The second classroom story describes one "turn" of a Roll and Build activity in a New Jersey preschool classroom serving a high percentage of dual-language learners. A partial transcript of verbal interactions is provided because we were unable to obtain parental permissions for the sharing of the video. The teacher has slightly adapted the Roll and Build activity for a small-group learning experience for seven children.

The teacher and children are sitting at a semicircular table. Two children are designated as rollers (they take turns), and they are also in charge of identifying the number represented on the die. Two children are builders, sharing a container of colorful foam blocks. Three children are "writers" and have markers and small whiteboards.

Teacher: (to child) You're going to roll? Where are my builders? My builders? Are you ready? (Pointing) Are you ready, builders? Okay. Writers. Are you ready, writers? Are you ready? (Children: Yeah!) Okay. Get your markers. Okay, J—, you're gonna roll.

J— rolls the die. Another child identifies the roll as "a six!" The teacher asks J—, "Can you count them?" She points as he accurately counts to six.

Teacher: Okay, writers . . . show me six.

One writer says, "I did a six." Another child says, "Yeah, that's a big one." Another says, "Good job."

One builder has already removed six blocks from the container, accurately counting out loud as she removes blocks from the container one by one. She ends with "six." The teacher says, "Okay. We have six. Builders?" The girl makes a structure with the six blocks she has chosen.

The other builder, meanwhile, has begun pulling blocks and counting. He takes two out and begins counting: "One. . . ." From that point, he counts accurately and ends by saying, "Six."

The teacher says, "Do you want to count them again, just to make sure?"

The boy points to each block, tagging each with the correct number name. When he says "six" but notices that he has a

	block left over, he quickly and silently moves it back into the container while another child says, "Seven."
Teacher:	Why'd you put that back?
Another child:	'Cause it's seven.
Teacher:	'Cause it's seven?
Child:	Yeah.
Teacher:	And what's the number?
Children:	Six!
Teacher:	Six. So did he use too many or too little?
Child:	Too many.
Teacher:	Too many. So you put one back (to builder). Are you done building? All right. Are you ready for the next part? All right, C—, do a nice roll.

In the coaching meeting after this lesson was observed, the teacher and master teacher (coach) reflected on the lesson and the video together. Based on evidence from the lesson, they noted that the children's learning goals had been met. Not only were the children engaged throughout, but they were able to identify and write numbers. (There were also opportunities to recognize various representations of numbers, such as numerals and pips on a die, to count and use one-to-one correspondence, and to engage in numerical comparison and reasoning.) The educators discussed possible extensions to the lesson, including incorporating geometry (identification of the 3-D shape of the blocks—cylinder, rectangular prism, cube) and data collection ("How many blocks of each shape are in your structure?").

8

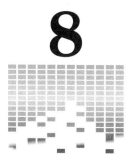

Evaluating Professional Development in Early Childhood Mathematics

Jessica E. Vick Whittaker and Bridget K. Hamre

Given clear evidence of the influence of children's early math skills on their later academic development, there is no doubt that to be effective, today's early childhood teachers need to know how to teach math and assess children's math learning. Unfortunately, we also know that few current teachers possess the necessary knowledge and skills (Copple, 2004; National Mathematics Advisory Panel, 2008), and few teacher preparation programs adequately cover early math learning and teaching in their preservice coursework (see Preston, Chapter 6; Ginsburg et al., 2006). The field has responded to these challenges by developing a growing number of professional development programs aimed at preparing teachers to support children's mathematical development and enhance the effectiveness of implementation of math curricula (e.g., Clements, Sarama, Spitler, Lange, & Wolfe, 2011; Ginsburg et al., 2006; Kinzie, Vick Whittaker, Kilday, & Williford, 2012; Preston, Campbell, Ginsburg, Sommer, & Moretti, 2005; Starkey, Klein, & Wakeley, 2004). These programs are often coupled with specific curricula and typically include a series of workshops (e.g., Ertle, Ginsburg, & Lewis, 2006; Hill & Ball, 2004; Hill, Rowan, & Ball, 2005). They are sometimes also combined with coaching (e.g., Starkey et al., 2004) and/or a technology-based training component (e.g., Clements, Sarama, Spitler et al., 2011; Kinzie et al., 2012). A few models have also been developed specifically for preservice teachers (Ginsburg et al., 2006; Preston et al., 2005).

Despite the clear need for these types of programs, rigorous evaluations of these professional development models are rare (National Research Council [NRC], 2009). When research does exist (e.g., Clements, Sarama, Spitler et al., 2011; Kinzie et al., 2012; Klein, Starkey, Sarama, Clements, & Iyer, 2008;

Starkey et al., 2004), professional development is often evaluated in a way that makes it impossible to disaggregate the effects of curriculum from professional development, leading to a lack of information about the specific elements of coursework, coaching, or workshops that may be most important for changing teacher practice and improving children's math learning. For example, studies of the Building Blocks curriculum and its associated professional development (workshops, coaching, online resources) have shown positive impacts on both teacher practice and children's math outcomes (Clements, Sarama, Spitler, et al., 2011). However, it is unclear which of the ingredients led to change in teacher practice. Thus practitioners and policy makers wanting to address issues related to early childhood teachers' competency in mathematics, but often with limited resources to do so, are left with little empirical data to drive decision making around program implementation or resource allocation.

In this chapter, we assert that significant progress in mathematics education for young children will not occur unless we are able to consistently develop and deliver high-quality pre- and in-service coursework and professional development. The best route to developing programs for teachers that are effective, affordable, and scalable is to engage in a systematic study of early childhood mathematics professional development. This will require the articulation of specific theories of change. Theories of change provide a clear road map for reaching goals by outlining the specific ways in which professional development experiences can improve teacher knowledge, beliefs, and skills, and the ways in which these, in turn, provide a positive impact on children's mathematics knowledge and skills.

We first discuss one broad theory of change, the Intentional Teaching Framework (Hamre, Downer et al., 2012), that describes a process through which professional development may affect teachers' classroom practice. We then use this model to guide a discussion of how it could be used to evaluate early childhood math professional development and the types of measures available for this evaluation. We then highlight examples of evaluation work in early childhood math professional development, drawing from both preservice and in-service programs. For each example, we describe the professional development being implemented, outline how it aligns with the Intentional Teaching Framework, and describe how the professional development is being evaluated. We conclude with a discussion of the impact of learner characteristics on the translation of professional development to practice and the implications for future research on the evaluation of early childhood math professional development.

DEVELOPING AND TESTING PROFESSIONAL DEVELOPMENT THEORIES OF CHANGE

Although different professional development approaches may articulate varying theories of change, all should include components related to children's development, teacher beliefs, knowledge, and/or practice, and the specific elements of professional development that are designed to promote change. For many years, we lacked the conceptual and measurement tools necessary to test these theories of change, but as suggested throughout this volume, significant progress has been made in these areas.

Now more than ever, we have an understanding of what young children know and need to know about math as well as the ways in which math knowledge and skills develop in the early years (Clements & Sarama, 2009; NRC, 2009). There has also been significant progress in our ability to assess children's early math knowledge and skills. For example, the Tools for Early Assessment in Math (TEAM; Clements, Sarama, & Wolfe, 2011), an assessment of children's number and geometric/spatial competencies, is based on research on the developmental progression of skills in these areas and was created using observations of children's behavior, replicated with data found from these observations, and then refined through clinical interviews, focused observations, and statistical analysis (Clements, Sarama, & Liu, 2008).

Researchers have also started identifying teacher knowledge, interactions, and strategies that are related to children's growth in math content domains. Effective teaching requires a combination of knowledge of mathematics content and young children's math development within these areas, explicit use of strategies and materials, high-fidelity implementation of an appropriate math curriculum, and instructionally supportive interactions (see Copley, Chapter 4; Clements & Sarama, 2007; Mashburn et al., 2008; NRC, 2009; Sarama & Clements, 2009). The National Association for the Education of Young Children (NAEYC) and the National Council of Teachers of Mathematics' (NCTM) position statement (2002) outlined 10 research-based recommendations to guide mathematics instruction including, using a curriculum and teaching practices that strengthen children's math and problem solving skills, integrating math with other activities, using ongoing assessment, and enhancing children's natural interest in math. Despite a better understanding of the general recommended practices that are related to children's mathematics skills, there remains a need for greater precision in our understanding of the very specific types of teacher practices that are associated with specific elements of mathematical knowledge and skill.

Improvement in children's math knowledge and skills depends on the alignment of three factors: 1) a menu of professional development inputs to teachers with 2) teacher knowledge, skills, and classroom practices that produce 3) skill and knowledge gains in math. Just as there has been a rigorous process to articulate the math knowledge and skills that we want children to exhibit in early childhood, we need a similar process to articulate and assess the very specific teacher practices related to gains in children's math attitudes, approaches to learning, and knowledge and skill development. Observations of teachers' interactions and teaching strategies that lead to children's conceptual understanding in specific content areas can also lead to teacher practice targets that are refined through interviews with teachers, observations, and statistical analysis. Professional development should thus be focused on improving the teacher interactions and strategies that have been proven to enhance children's learning. This professional development should then be tested to ensure that it is having its desired impact on teachers' classroom interactions and strategies and, ultimately, child outcomes.

There is limited information on the theory of change underlying early childhood math professional development. This makes it difficult to develop clearly articulated targets of professional development that are linked to teacher practice. For example, the NAEYC and NCTM position statement (2010) outlines six features that should be incorporated into in-service professional development: 1) teacher networking or study groups, 2) sustained intensive programs, 3) collective participation of all staff who work in similar settings, 4) content focused on both what and how to teach, 5) active learning techniques, and 6) professional development as part of a coherent program of teacher learning. However, these features are not explicitly linked to the NAEYC and NCTM recommendations around classroom practice.

Although there is growing research on the effectiveness of specific early childhood professional development programs (see Zaslow, Chapter 5), there remains a lack of understanding around the process by which early childhood professional development programs influence specific elements of teaching and learning (Sheridan, Edwards, Marvin, & Knoche, 2009). Most of the recommendations are general and fail to clearly articulate the very specific types of knowledge and skills that teachers need and the ways in which particular types of professional development may foster these. For example, there is general agreement that coaching can be an effective form of professional development. But why? What is it that coaches do in interacting with teachers that promotes positive changes in practice? Until we can more clearly identify the specific elements of professional development that work, we will be limited in our ability to consistently and efficiently build and scale up effective models. To address some of these issues Hamre, Downer, Jamil, and Pianta (2012) have articulated a theory of change by which they assert that professional development may have positive effects. In the next section, we describe this conceptualization and provide examples, where available, of the potential efficacy of this model in studying early childhood math professional development.

INTENTIONAL TEACHING: A CONCEPTUAL MODEL FOR TEACHER PROFESSIONAL DEVELOPMENT

Hamre, Downer, and colleagues (2012) hypothesize that professional development may have positive effects by helping to make teachers more intentional in their classroom practice. Intentionality is a word that is used a lot in education, but Hamre, Downer, and colleagues (2012) operationalized this practice by articulating four components of intentional teaching: knowing, seeing, doing, and reflecting. In conceptualizing this theory of change and operationally defining the associated components, these researchers have developed theoretically based, measurable targets for professional development that are known to relate to child outcomes across content areas. Figure 8.1 depicts the components of intentional teaching as they relate specifically to mathematics. In what follows, we briefly describe each component of the model. For each component, we comment on measures that exist in relation to each in the field of early childhood and specifically math, where available.

Knowing

Research and practice have long recognized the importance of teacher knowledge in guiding effective and intentional practice (Shulman, 1987). In their

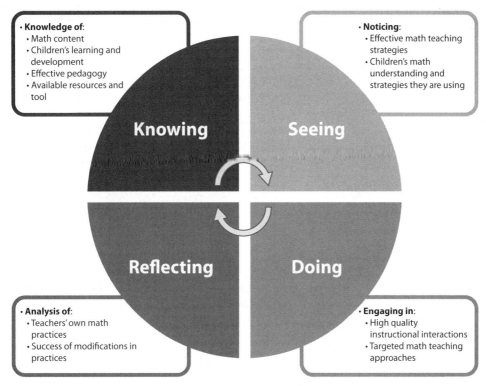

Figure 8.1. Intentional Teaching Framework. (Source: Hamre, Downer, Jamil, & Pianta, 2012.)

conceptualization of "knowing," Hamre, Downer, and colleagues (2012) include knowledge of child development, knowledge of a range of effective teaching practices, knowledge of individual children's needs, content knowledge, and knowledge of self. In the area of early childhood mathematics, NAEYC and NCTM (2010) have outlined five critical areas of knowledge that early childhood educators must have to be effective in teaching mathematics to young children: knowledge of 1) math content they will be teaching, 2) children's learning and development, 3) effective math pedagogy, 4) effective means for assessing children's development, and 5) resources and tools available for teaching early childhood mathematics. Teachers' knowledge of mathematics is central to their ability to effectively use instructional materials, assess students' progress, and make judgments about how best to scaffold children's learning (Ball, Hill, & Bass, 2005; NRC, 2009). And yet almost no empirical research exists that examines early childhood teachers' math knowledge (see Hyson & Woods, Chapter 2 for further discussion).

However, there are measures that have been developed to assess elementary school teachers' knowledge for teaching mathematics. Specifically, Hill and colleagues' research has identified knowledge for teaching—defined as "mathematical knowledge used to carry out the work of teaching mathematics" (2005, p. 373)—as a key factor that predicts both the quality of teachers' instruction (Ball, 1991) and student math achievement in early elementary school (Hill et al., 2008; Hill et al., 2005). In noticing the lack of available instruments to assess mathematical knowledge for teaching, they developed the Mathematics Knowledge for Teaching (MKT) Measures to fill this gap. The measure contains the domains of number, operations, and geometry. Items include questions related to scenarios that teachers face in the classroom, and the measure assesses two types of knowledge: 1) common knowledge of mathematics that all adults should have and 2) specialized knowledge that only teachers would need to know (Ball et al., 2005).

Because this measure was developed for elementary teachers, not all items may be applicable to children in prekindergarten. However, both types of knowledge assessed by this instrument are included in the NAEYC and NCTM's (2010) critical knowledge areas for early childhood educators and are associated with quality teaching and student learning (Ball, 1991; Hill et al., 2005). This measure could serve as a template for developing teacher knowledge measures that are applicable for early childhood teachers. In fact, recent progress has been made in this area with the development of a teacher interview to assess early childhood teachers' pedagogical content knowledge (McCray & Chen, 2012). It is important that work in this area continues, as measures of teacher knowledge are critical in evaluating whether professional development produces teacher mathematical knowledge known to be important for classroom practice.

Seeing

The second component of the Intentional Teaching Framework is teachers' ability to observe and identify effective and ineffective teacher–child interactions, both in their own and in other teachers' practices (Hamre, Downer et al., 2012). Social learning theory (Bandura, 1986) and dynamic memory theory (Schank, 1982) suggest that people learn in part by observing and developing scripts for how to behave in different situations. This work suggests that teachers should learn about how to teach from watching examples of teaching—an idea that has been validated by work on "teacher noticing" (e.g., Van Es & Sherin, 2002). Observing other teachers' instruction is a common training technique for both preservice and in-service teachers (Anderson, Barksdale, & Hite, 2005; Borko, Jacobs, Eiteljorg, & Pittman, 2008; Hamre, Pianta et al., 2012; Hatch & Grossman, 2009; Santagata, Zannoni, & Stigler, 2007). Research also suggests that the practice of having teachers observe and reflect on their *own* teaching can lead to improvements in the quality of teacher-child interactions and instruction (Dickinson & Caswell, 2007; Pianta, Mashburn, Downer, Hamre, & Justice, 2008). However, there have been few efforts to explicitly understand how best to develop teachers' ability to be good observers of both their own and other teachers' classroom practice and how to define and measure this skill.

There have been some recent attempts in the field of secondary mathematics education to define and measure observations that could be useful models for the field of early childhood mathematics professional development. Star and Strickland (2008) designed a measure to explicitly assess whether a preservice course for secondary mathematics teachers improved their ability to notice and interpret salient events related to mathematics instruction in classroom lessons. In the fall and spring, teachers viewed full-length videos of an eighth-grade math period and completed an assessment designed to measure the quantity and type of classroom events that teachers noticed. They found that teachers made significant gains in their observation skills. This assessment was specifically developed for use with secondary preservice mathematics teachers. We know of no such instruments designed for use with early childhood preservice mathematics teachers; however, there is currently a measure under development designed to evaluate early childhood teachers' skill in observing overall (versus content specific) classroom interactions.

Video Assessment of Interactions in Learning (VAIL; see Jamil, Sabol, Hamre, & Pianta, 2012) assesses teachers' ability to detect effective classroom interactions by having them watch a series of videos of other teachers' classrooms. VAIL uses the Classroom Assessment Scoring System® (CLASS™; Pianta, La Paro, & Hamre, 2008) as a guiding framework, and teachers are asked to identify effective interactions that the teacher in the video is using. In a study of 270 preschool teachers, Jamil and colleagues (2012) found that early childhood teachers who could accurately detect effective interactions had more years of education and displayed more effective instructional interactions with students in their classrooms. And most important, VAIL appears to help unpack the effects of professional development on teachers' practice. Teachers who took a course on effective interactions scored higher on VAIL than did teachers in the control group, and VAIL scores were shown to mediate the impact of the course on changes in teachers' observed practice (Hamre, Pianta et al., 2012). In this example, CLASS is used as a framework for teachers to use to interpret classroom behavior. This could be more challenging in the field of early childhood mathematics, where there are many complexities around asking teachers "what to look for" and where teachers' own lack of mathematical knowledge may hinder their ability to detect appropriate strategies.

It is not only important for teachers to be able to notice and identify effective classroom strategies that promote early childhood mathematics, but it is also important that teachers be able to observe and identify the strategies that children are using and be able to interpret their mathematics understandings. Some early childhood math researchers have suggested that the use of video is a useful approach for helping teachers to develop these types of observation skills, noting that video can show examples of children's behavior and thinking and can be played multiple times to provide opportunities for analysis (see Preston, Chapter 6; Ertle et al., 2008). Measuring development in the area of teachers noticing both teachers' strategies to support early mathematics learning and children's understandings could move the field forward in better understanding how to support these types of skills and how they relate to children's outcomes.

Doing

The third component of Hamre, Downer, and colleagues' (2012) model is teachers' ability to engage in effective teaching. Citing the research on experiential learning (e.g., Kolb, Boyatzis, & Mainemelis, 2000) and expertise (e.g., Ericsson & Charness, 1994), they suggest that learning occurs through practice. The idea that practice is important in skills development is embodied in other fields (e.g., medicine, cosmetology, aviation), where training involves scaffolded apprenticeships and individuals are only allowed to practice independently once they have passed a performance assessment. Historically, the field of education has struggled to articulate a set of classroom practices linked to student learning that can and should be assessed as skill targets for teacher candidates, and to provide ongoing feedback and monitoring to in-service teachers (Zaslow, Tout, Halle, Whittaker, & Lavelle, 2010). Recent progress in this area has included a call for high-quality clinical practice to be included in preservice teacher educational programs (National Council for Accreditation of Teacher Education, 2010) and the development and piloting of a national assessment of preservice teachers' readiness to teach (Newton, 2010).

In the field of early childhood mathematics, several measures have been used to assess teachers' ability to "do," or engage in, effective teaching. Vick Whittaker Kinzie, Williford, and DeCoster (2013) have used CLASS (Pianta, La Paro et al., 2008) to assess the quality of teachers' instructional interactions (e.g., concept development, language modeling, and quality of feedback) during mathematics activities. Specific to early childhood mathematics, Sarama and Clements (2009) developed the Classroom Observation of Early Mathematics Environment and Teaching (COEMET), designed to assess the quality and quantity of mathematics teaching in early childhood. COEMET consists of two subscales: 1) Classroom Culture, which assesses teachers' general approach to teaching mathematics and personal attributes of teachers such as knowledge and enthusiasm, and 2) Specific Math Activities, which assesses constructs such as teaching approaches, teacher expectations, and support for children's mathematical thinking. These measures have good psychometric characteristics and are sensitive to change in the quality of teachers' instruction and interactions (Kilday & Kinzie, 2009; Pianta, Mashburn et al., 2008). These measures allow researchers to test whether their professional development has an impact on teachers' actual classroom practice. If measures of knowing and seeing are also included in evaluations, questions can be answered not only about the relation among these components (e.g., whether increases in teacher knowledge and their ability to perceive effective classroom instruction and interactions are related to effective teaching) but also about child outcomes.

Reflecting

The final component of Hamre, Downer, and colleagues' (2012) Intentional Teaching Framework is reflection. As with knowledge, reflection has long been recognized as central to effective and intentional teaching practice, dating back to early work by Dewey (1933). Hamre, Downer, and colleagues (2012) define reflection as the "active examination of teaching practice and the factors that may influence it" (p. 516) and suggest that the goal of reflection should be improvement in the effectiveness of teaching practices and the quality of interactions with students (Kottkamp, 1990). They conceptualize reflection as a cycle that includes observation, assessment, analysis of practice, and planning for change in practice. There is less empirical evidence on the link between the quality of teachers' reflection and their practice, but Hamre, Downer, and colleagues (2012) included this as a component in the model based on the hypothesis that teachers' analysis of their own teaching and the ability to plan for change are critical components in improving teachers' practice.

Several early childhood math professional development programs include an implicit focus on teacher reflection (e.g., Kinzie et al., 2012; Sorkin & Preston, 2010; Starkey et al., 2004). However, there are few measures available to assess the impact of professional development on teachers' ability to reflect. This is in part due to uncertainty about the best context (e.g., during coaching sessions or as part of in-service trainings), format (e.g., verbal or written), and content (e.g., teacher-child interactions, attitudes, and beliefs) for teacher reflection, and a lack of well validated measures designed to assess teachers' reflectivity. Jamil (2013) is working to fill this gap by developing a theoretically based measure of teacher reflection that could be used to assess the impact of professional development on the quality of teachers' reflection and is exploring whether teacher reflection is related to the quality of teacher-child interactions in the classroom. Measures such as this one will allow the developers of early childhood mathematics professional development to make their focus on teacher reflection more explicit and to evaluate the impact of the professional development on the quality of teacher reflection.

In the next section, we provide four examples of evaluation work in early childhood math professional development, drawing from both preservice and in-service programs (see Table 8.1 for a summary). For each example, we describe the professional development being implemented, outline how it aligns with components of the Intentional Teaching Framework (know, see, do, and reflect), and describe how the professional development is being evaluated. The evaluation work presented here was conducted by researchers. However, we have highlighted important policy and practice implications that can be learned from each example.

EXAMPLES OF EVALUATION WORK IN EARLY CHILDHOOD MATH PROFESSIONAL DEVELOPMENT

An Evaluation of California's Mathematics Professional Development Institutes

Hill and Ball's research (e.g., Ball et al., 2005; Hill & Ball, 2004; Hill et al., 2005) has focused on testing the links among teachers' math professional development, their professional math knowledge, and students' achievement. Their program of research, as already described, has included the development and testing of a measure designed to assess teachers' mathematical knowledge for teaching.

To evaluate the impact of professional development on teacher knowledge, Hill and Ball (2004) evaluated California's Mathematics Professional Development Institutes to determine whether they had an impact on teachers' knowledge for teaching mathematics. Elementary school teachers ($n = 398$) attended summer institutes between 1 and 3 weeks long, taught by mathematicians and mathematics educators. Teachers were administered a pre- and post-test measure of elementary number concepts and operations developed by the researchers to assess mathematical knowledge for teaching. Although their analysis was limited by the fact that the survey was still under development,

Table 8.1. Summary of examples of early childhood mathematics professional development evaluations

Early childhood math professional development program	Professional development goals	Professional development components
An Evaluation of California's Mathematics Professional Development Institutes (Hill & Ball, 2004)	**KNOW** Determine impact of California's Mathematical Professional Development Institutes on teachers' knowledge for teaching mathematics	Summer institutes between 1 and 3 weeks long taught by mathematicians and math educators
Video Interactions for Teaching and Learning (VITAL; Sorkin & Preston, 2010)	**SEE** Train early childhood preservice teachers on how to observe and interpret children's behavior in order to prepare them for classroom teaching	Video library of topics related to young children's mathematical thinking and learning Video viewing tool called Media Thread that allows preservice teachers to edit and annotate videos A "workspace" where students complete multimedia essay assignments that require them to interpret videos and attach video clips as examples of support for their hypotheses about teaching and learning
Technology-Enhanced, Research-Based, Instruction, Assessment, and Professional Development (TRIAD; Clements & Sarama, 2011)	**KNOW** Increase teachers' mathematical content knowledge Improve teachers' understanding of children's developmental progression of math thinking and learning **DO** Improve the quality of instructional tasks and teaching strategies that help children move forward in their progression	Series of professional development workshops across the year Web-based resources on children's learning trajectories including videos of children's thinking and recommended practices Coaching via one-to-one consultation to help teachers with planning for and reflecting on their classroom
MyTeachingPartner-Math/Science (MTP-M/S; Kinzie et al., 2012; Vick Whittaker et al., 2013)	**KNOW** Increase teachers' content and pedagogical knowledge **DO** Promote high-quality instructional interactions in math and science **REFLECT** Provide opportunities for self- and peer reflection	Online supports: • Video demonstrations of each activity • Teaching tips on best pedagogy and student misconceptions • CLASS Quality Teaching Library In-person workshops with video observation and reflection on practice

they found that longer institutes—and opportunities within the institutes to explore and link alternative representations, provide and interpret explanations, and make connections among ideas—led to gains in teachers' content knowledge for teaching mathematics.

Hill and Ball have also tested the association between teachers' knowledge of mathematics for teaching and students' math knowledge and skills (Ball et al., 2005). As part of the Study of Instructional Improvement, first- and third-grade teachers ($n = 700$) completed items from their MKT measure, and students' math skills were assessed at the beginning and end of the year using a standardized measure. They found that teachers' performance on the MKT measure significantly predicted students' gain scores.

Hill and Ball's research on teachers' math knowledge for teaching begins to provide an evidence base about what early elementary school teachers know about mathematics and how that knowledge is associated with teachers' classroom practice and children's mathematics knowledge and skills. Furthermore, they have started identifying key components of professional development that can positively influence teacher knowledge. Although their work has included elementary and middle school teachers and children, it suggests that an important next step in the evaluation of early childhood mathematics professional development is assessing the impact of these experiences on teachers' content and pedagogical knowledge and implementing further testing about the associations between this knowledge and children's math skills and abilities.

The National Research Council (2009) suggested that an important way to increase teachers' mathematical knowledge is through preservice coursework that helps them deeply understand mathematical concepts. Preston and colleagues (2005) have been working to develop and evaluate the impact of preservice courses designed to increase teacher knowledge and their ability to effectively observe and analyze children's math capabilities.

Video Interactions for Teaching and Learning

Video Interactions for Teaching and Learning (VITAL) is a web-based application that was developed by Columbia University's Center for New Media Teaching and Learning to train early childhood preservice teachers on how to observe and interpret children's behavior in order to prepare them for classroom teaching (http://ccnmtl.columbia.edu/vital/nsf; Sorkin & Preston, 2010). VITAL includes 1) a video library of topics related to young children's mathematical thinking and learning, 2) a video viewing tool called Media Thread that allows preservice teachers to edit and annotate videos, and 3) an online workspace where students complete multimedia essay assignments that require them to interpret videos and attach video clips as examples of support for their hypotheses about teaching and learning (Preston et al., 2005; Sorkin & Preston, 2010).

Preservice teachers' ability to effectively observe children's thinking and learning is assessed in multiple ways throughout the course. First, preservice teachers' responses to their multimedia essay assignments are read by the instructor and peers, and feedback is given by the instructor. Preservice teachers also participate in a series of video lessons that require them to respond to specific video clips and questions. As the preservice teacher watches a clinical interview with a child, the video pauses, and the student is prompted to make a specific assessment about a child's understanding of a concept. For example, "What did you observe about how the child got the answer to this problem?" (Preston et al., 2005). Finally, preservice teachers complete a final project that involves designing a mathematical lesson or activity, trying it out with a child, and interviewing the child to find out what he or she learned. The preservice teacher records this interaction and writes a research paper that requires him or her to analyze his or her own teaching (Sorkin & Preston, 2010).

Initial evaluation of VITAL's work is assessing whether it helps preservice teachers to be better observers of children's knowledge and skills and whether it improves their ability to use the observations to support a theory about children's mathematical thinking. In a qualitative analysis of eight preservice teachers' essays over one semester (eight essays per preservice teacher), Preston and colleagues (2005) developed codes to categorize how students used observations of children's mathematical thinking. They found that students used video observations in their essays as evidence, analysis, and to make connections. The VITAL training program provides a ripe context for the study of preservice teachers' skill in detecting and identifying children's behavior and thinking.

Work using VITAL provides initial evidence that early childhood math professional development can be effective in improving preservice teachers' ability to better attend to important aspects of children's language, strategy use, and behavior in interpreting children's early math understandings. It also suggests that video can be a useful tool in helping to develop this important skill in future educators. As Preston highlighted (Chapter 6), new technology has made it much easier for teachers to create, upload, and review videos of themselves and others teaching. Future evaluation work in the area of early childhood mathematics professional development should continue to assess how best to foster both pre- and in-service teachers' observation and detection skills (of both effective teaching practices and children's mathematics knowledge and skills). Ideally, this work will progress using a variety of methodologies, including the use of quantitative measures of teachers' observational skills (e.g., Jamil et al., 2012) that will allow researchers to test the impact of coursework, demonstrate the ways in which these skills transfer to improved classroom teaching, and ultimately provide a clear description of a set of skills that can become targets for other interventions.

In the next section, we describe a program of research designed to evaluate the impacts of an in-service program of professional development on teachers' knowledge, skills, practice, and child outcomes.

Technology-Enhanced, Research-Based, Instruction, Assessment, and Professional Development

Technology-Enhanced, Research-Based, Instruction, Assessment, and Professional Development (TRIAD) was developed to support in-service early childhood teachers' implementation of the preschool mathematics Building Blocks curriculum by improving their mathematical content knowledge, understanding of children's developmental progression of math thinking and learning, and instructional tasks and teaching strategies that help children move forward in their progression (Clements & Sarama, 2007, 2011; Clements, Sarama, Spitler et al., 2011; Sarama, Clements, Starkey, Klein, & Wakeley, 2008). The TRIAD model includes multiple supports for teachers to help with this improvement. The first main component is a series of professional development workshops across the school year that include an introduction to children's learning trajectories in math, mathematics content and how best to teach it, and hands-on experiences around the curriculum's activities. Teachers are given access to a web-based application, Building Blocks Learning Trajectories, which provides access to the learning trajectories, videos of children's progression in thinking, and video exemplars of recommended practice to illustrate high-quality teaching and implementation of the curriculum. The second component of the professional development includes coaching. Coaches visit teachers once a month and provide one-to-one consultation to help teachers with planning for and reflecting on their classroom practice and collaborative problem solving.

Clements and Sarama have conducted multiple evaluations of the Building Blocks curriculum and associated professional development. Most recently, they conducted a randomized trial in 42 schools and 106 prekindergarten classrooms serving low-resource communities to examine the impact of the curriculum and associated professional development on teachers' fidelity of implementation, the quality of the mathematics environment and teaching, and child outcomes (Clements, Sarama, Spitler, et al., 2011). They further examined whether there was an indirect effect of the curriculum and professional development on child outcomes through their influence on the quality of teachers' instruction and the mathematics environment. Schools were assigned to either the intervention group or a business-as-usual comparison group. In order to assess teachers' ability to "do," or engage in, effective teaching, the researchers used the Fidelity of Implementation measure (Clements, Sarama, Spitler, et al., 2011) and the COEMET (Sarama & Clements, 2009).

Teachers implemented the Building Blocks curriculum with adequate fidelity, and Building Blocks classrooms provided richer environments for mathematics learning (Clements, Sarama, Spitler et al., 2011). Compared with control teachers, Building Blocks teachers offered richer classroom environments for mathematics, engaged in more math activities, spent more total time on math, and had more computers available for students to use. Related to children's outcomes, children in Building Blocks classrooms, compared with those in control classrooms, scored significantly higher on a measure of core mathematical abilities (REMA; Clements et al., 2008). Finally, the researchers found that the impacts of the intervention on children's math development were the result, in part, of the improved mathematics environments (i.e., number of computers available for children to use, number of specific math activities, and classroom culture). Importantly, the researchers found that the professional development was equally as effective in classes serving low- or mixed-income families or in schools serving a higher or lower proportion of children with limited English proficiency.

The work described here provides an example of how the impacts of a curriculum and professional development on teachers' classroom interactions and instruction and children's math learning was developed and rigorously tested. Through this evaluation, Clements and Sarama were able to unpack some of the mechanisms around the influence of the professional development by including not only measures of child outcomes but also assessments of environmental and instructional quality. In order to unpack the mechanisms around how early childhood professional development can ultimately have an impact on child outcomes, it is important to include potential mediating variables, such as the quality of the instructional environment and specific mathematics teaching practices.

However, the results of this study leave questions about which part of the professional development has had an impact on the quality of the mathematics environment. It is unclear whether teachers' practice would have improved solely as a result of implementing the curriculum or whether the professional development supports offered an advantage above and beyond the curriculum.

In the next section, we describe the evaluation of an early childhood math curriculum and system of professional development supports that starts to disentangle the impacts of curriculum from professional development.

MyTeachingPartner-Math/Science

MyTeachingPartner-Math/Science (MTP-M/S; Kinzie, Vick Whittaker, McGuire, Lee, & Kilday, 2013) consists of early childhood math and science curricula paired with professional development supports for in-service preschool teachers. The teaching supports are intended to promote high-quality instructional interactions in the areas of math and science and high-fidelity implementation of the MTP-M/S curricula. MTP-M/S includes a blended model of professional development with both online supports and teacher workshops.

Based on research suggesting the importance of visual exemplars in encouraging teachers' ability to describe specific applications of targeted teaching skills (Moreno & Ortegano-Layne, 2008), Kinzie and colleagues (2012) developed several web-based supports including 1) more than 130 2- to 3-minute video demonstrations of high-fidelity curricular implementation that also embody the characteristics of high-quality teacher–child interactions (see Figure 8.2); 2) weekly video-based 5-Minute Quality Teaching Challenges, featuring one of the activities teachers will implement that week (see Figure 8.3); and 3) links to a Quality Teaching Library offering 150 video examples across many instructional settings, formats, and content areas.

Figure 8.2. Demonstration video. (From Kinzie, M., Pianta, R.C., Whittaker, J., Foss, M.J., Pan, E., Lee, Y., Williford, A.P., & Thomas, J.B. [2010]. MyTeachingPartner Math/Science demonstration video. Charlottesville, VA: MyTeachingPartner Math/Science; reprinted by permission.)

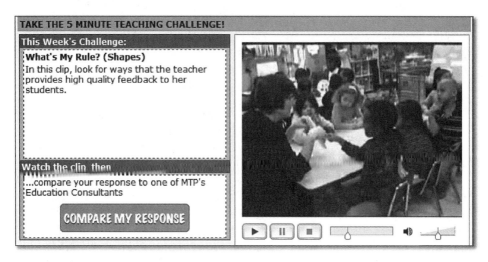

Figure 8.3. 5-Minute Quality Teaching Challenge. (From Kinzie, M., Pianta, R.C., Whittaker, J., Foss, M.J., Pan, E., Lee, Y., Williford, A.P., & Thomas, J.B. [2010]. MyTeachingPartner Math/Science demonstration video. Charlottesville, VA: MyTeachingPartner Math/Science; reprinted by permission.)

Eight professional development workshops guide teacher processing of these supports and explicitly offer opportunities for self- and peer reflection. Reflection on the practice of other teachers facing the same teaching challenges (with similar students and equivalent learning goals) was encouraged during the workshops. Between workshops, teachers completed reflective assignments including review of online supports, video analysis of their own teaching with a focus on strategies presented at the workshops, and peer-debrief sessions.

Kinzie and colleagues (2012; Vick Whittaker et al., 2013) have conducted a small field trial evaluating the effects of the MTP-M/S curricula and professional development supports on teachers' fidelity to recommended practices in teaching math and science, adherence to the curricula, quality of math and science instructional practices, and children's math and science outcomes. To try to disaggregate the impacts of the curricula from the professional development, they randomly assigned teachers to one of three conditions representing variations in curricular access and professional development support. Teachers in the control group implemented the district's existing math and science curricula ($n = 9$). Teachers in the MTP-M/S Basic group ($n = 15$) received the math and science curricula, plus the teaching materials needed to implement the activities. Teachers in the MTP-M/S Plus group ($n = 11$) received the same MTP-M/S curricula and materials as the Basic group and also received the online supports and workshops. Results suggested that teachers in the Plus and Basic groups outperformed teachers in the control group on measures of Instructional Support (Pianta, La Paro et al., 2008) as well as fidelity to recommended practices in teaching math and science. Although teachers in the Basic and Plus groups adhered equally well to the curricula, there was greater variability in the Basic group. In other words, there was a wider range in adherence to the curricula for Basic teachers, with some in this group who did not implement them as intended.

With regards to child outcomes, children of teachers in the Plus ($n = 146$) and Basic ($n = 182$) groups scored significantly higher than those in the control group (n = 116) on a measure of geometry and measurement (derivative of the TEAM; Clements, Sarama, & Wolfe, 2011), and children of teachers in the Plus group outperformed children in the control and Basic groups on a measure of number sense and place value (Kinzie, Vick Whittaker, Williford et al., 2013).

The results of this study provide an example of planned variation (comparing a group of teachers who receive curricula alone to a group who receives curricula and professional development supports) and a method for beginning to understand how individual components of professional development are related to teacher practices and child outcomes. In combination, these results suggest that the MTP-M/S curricula alone had positive impacts on teacher practice and child outcomes. This has important implications given that the purchase and adoption of a curriculum may be a less expensive alternative to providing professional development supports. However, in the case of MTP-M/S, the curricula were designed in such a way that professional development supports were embedded throughout the written curricula (e.g., recommendations for language for teachers to model and elicit, adaptations for differentiated instruction, suggested extensions). In addition, the results do suggest the greatest impact on child outcomes for teachers who were provided with both the curricula and professional development supports.

Taken together, these highlighted examples provide a snapshot of innovative evaluations of the impact of early childhood math professional development. Each of the early childhood mathematics professional development programs targeted improvement in teachers' knowledge, observation skills, effective teaching, reflection, or some combination of these. We are learning more to answer the question "What works?" However, in the field of early childhood mathematics, we have little research that answers the next question: "What works for whom?"

THE ROLE OF TEACHER CHARACTERISTICS IN THE TRANSLATION OF PROFESSIONAL DEVELOPMENT INTO PRACTICE

Even when professional development has been rigorously evaluated and found to be effective in influencing teachers' ability to know, see, do, and reflect, some teachers may benefit more than others. Teachers differ in their ability to take advantage of professional development, in part due to personal and professional characteristics such as background education, emotional regulatory capabilities (Raver, Blair, & Li-Grining, 2012), and readiness for change (Dall'Alba & Sandberg, 2006).

Some studies suggest that teachers' beliefs about math may play a particularly important role in their classroom practice (e.g., Pajares, 1992; Santagata, 2005; for further discussion of beliefs about early childhood mathematics education, see Hyson & Woods, Chapter 2). Generally, early childhood teachers do not place a high importance on teaching math, feel anxious about teaching the subject, and do not feel competent in their ability to effectively teach it (Blevins-Knabe, Austin, Musun, Eddy, & Jones, 2000; Hart, 2002; Yesil-Dagli, Lake, & Jones, 2010). Ginsburg and colleagues (2006) argue that it is critical to directly address teachers' beliefs around mathematics within the context of professional development.

In order to understand the association between teacher beliefs and engagement in professional development and subsequent practice, there must be reliable methods and measures of teacher beliefs. Hart (2002) developed the Mathematics Belief Instrument (MBI) for preservice elementary teachers, which assesses individuals' math teaching philosophy, beliefs about teaching and learning mathematics, and self-efficacy. Research has suggested an association among preservice teachers' participation in an education course designed to develop preservice teachers' understanding of mathematics pedagogy, children's math learning, a positive approach to teaching math, and their score on the MBI (Hart, 2002; Wilkins & Brand, 2004). Ginsburg and colleagues (2006) have also developed a questionnaire to evaluate the impact of a preservice course on students' beliefs about early childhood mathematics education and found that students developed a better understanding of appropriate mathematical content, increased the amount of time they thought should be spent on early childhood mathematics, and had a better understanding of what young children are capable of learning. This research suggests that professional development, particularly preservice courses, can be effective in improving teacher beliefs, but it is unclear how changes in these beliefs are associated with changes in teachers' knowledge, skills, and practice.

IMPLICATIONS AND DIRECTIONS FOR FUTURE RESEARCH

In their report on mathematics learning in early childhood, the National Research Council (2009) recommended that evaluations of teacher preparation move beyond identifying whether or not teachers have a bachelor's degree to focus on the content and quality of teacher professional development. In the years since the report was issued, we have seen some shift in focus on identifying and understanding the process through which professional development has an impact on teachers' mathematical knowledge and practice. In particular, researchers have responded to the call to be more precise and include more detail about the context of and content included in their professional development (Martinez-Beck & Zaslow, 2006) as well as the need for more rigorous evaluation of that professional development (Zaslow et al., 2010). For example, in the evaluations highlighted in this chapter, information was available on the content included in the trainings, the number of training activities provided, and whether the training included coursework, workshops, coaching, or some combination of these. We also highlighted several experimental studies that included not only child outcomes but also some measure of teacher knowledge (Hill & Ball, 2004), skill (Sorkin & Preston, 2010), and practice (Clements & Sarama, 2008; Kinzie et al., 2012).

However, there are several remaining questions around the underlying theory of change associated with many early childhood math professional development programs; the links between teacher knowledge, skills, and practice, and child outcomes; and how best to evaluate the process through which professional development has its effects. In the sections that follow, we outline several recommendations for future research in the evaluation of early childhood mathematics professional development.

Develop and Test a Theory of Change by Which Early Childhood Mathematics Professional Development Has Its Effects

As outlined in this chapter, progress has been made on the conceptual and measurement work necessary to articulate and evaluate a theory of change that includes components related to children's development, teacher beliefs, knowledge, and/or practice as well as specific elements of professional development designed to promote change. We have introduced the intentional teaching model as a framework that could be used to study the mechanisms through which professional development has its effects. Using such a model, and testing each individual component, could lead to a clearer understanding of the types of professional development that have an impact on specific teacher practices and strategies that, in turn, lead to growth in specific areas of children's math development. Systematic evaluation of these components can provide further information about the relative importance of each component and the strength of their association with child outcomes. In addition, if multiple components are included in professional development, we can start to better understand how they work together as a system to produce change in teacher practice.

Develop Measures of Early Childhood Teachers' Math Knowledge, Perception, and Reflection

There are still many unanswered questions related to the types of knowledge that teachers need for effective teaching (see Ball, Thames, & Phelps, 2008), how best to assess this knowledge, and whether some are more predictive of child outcomes than others. In their 2008 report, the National Mathematics Advisory Panel wrote,

> Research on the relationship between teachers' mathematical knowledge and students' achievement confirms the importance of teachers' content knowledge. It is self-evident that teachers cannot teach what they do not know. However, because most studies have relied on proxies for teachers' mathematical knowledge (such as teacher certification or courses taken), existing research does not reveal the specific mathematical knowledge and instructional skill needed for effective teaching. (p. xxi)

This remains true, especially in early childhood where there are few direct assessments of teachers' mathematical knowledge (see McCray & Chen, 2012 for an exception). Researchers have suggested that self-report measures are not adequate to assess this knowledge and that new tools and procedures are needed to reliably assess the different types of knowledge needed to teach mathematics (Hill et al., 2008). More information is also needed on the types and content of both pre- and in-service professional development that increase this knowledge.

As described in this chapter, there has been recent measurement work in the areas of teachers' ability to notice effective classroom interactions and children's skills and abilities (Jamil et al., 2012; Star & Strickland, 2008). Lessons learned from the development of these measures could help to inform the development of a measure to specifically target teachers' observation and detection skills in the area of early childhood mathematics. In the development of this type of measure, it will be important to think about both teachers' ability to detect their own mathematics teaching strategies and also their ability to notice and interpret children's mathematics knowledge and skills.

Similarly, several of the programs described included reflection as an implicit component of the professional development. However, this construct is often imprecisely and variously defined, and we found no measures used to assess the impact of the early childhood mathematics professional development on the quality of teachers' reflection. As new measures are developed that assess this construct, evaluations of early childhood math professional development should consider including them.

A final issue related to the development of measures is that the majority of available assessments are used to evaluate early childhood mathematics professional development were developed by researchers and intended for use within the content of research to determine program impacts. The field has very few assessments that were developed for the purposes of planning and monitoring of teachers' progress or that were designed for use by practitioners. It is possible that some of the researcher-developed tools could be used for planning, monitoring, and improvement purposes, but assessments must be carefully evaluated to determine whether they can be reliably and validly used for those purposes.

Conduct Evaluations of Preservice Mathematics Training Programs

We still know very little about the efficacy of preservice mathematics programs. Preservice teacher preparation in the area of math is highly variable (National Research Council, 2009), but most research suggests that there is very little math coursework required of preservice teachers and that there is considerably less preservice training in math than there is in literacy (see Ginsburg et al., 2006; Hyson & Woods, Chapter 2; Lobman, Ryan, & McLaughlin, 2005). In fact, many early childhood teacher preparation programs require only one course in math, which often does not focus on early childhood (Ginsburg et al., 2006), and some require none at all. In evaluating this lack of math content in preservice training, Ginsburg, Lee, and Boyd (2008) have put forth recommendations around what should be included in preservice training, such as research on children's math learning, formative assessment, knowledge about children's math learning, exposure to various curricula and pedagogy, and analysis of video involving children's thinking.

Ginsburg and others (Ginsburg et al., 2006; Preston et al., 2005) have begun implementing a higher education preservice training program that embodies these recommendations. Evaluations of these courses have included measuring the impact of the course on students' beliefs about the importance of early childhood math education, children's early math abilities, and appropriate math pedagogy (Ginsburg et al., 2006). However, there is still a lack of research on how and whether teachers are able to include the knowledge and skills learned during preservice training with their practice once they enter the classroom.

Use Planned Variation Studies and
Small-Scale Experimental Studies to Identify
Effective Professional Development Components

Much of the evaluation of in-service early childhood math professional development includes both the delivery of a curriculum and its associated professional development. There remain unanswered questions about whether teachers' practice improves solely as a result of implementing a curriculum or whether professional development supports offer an advantage above and beyond the curriculum. Similarly, early childhood math professional development often includes multiple components (e.g., workshops, coaching, access to online supports), and it is unclear which of these professional development components are responsible for the improvements in teacher practice. Therefore, even where we have math curricula and professional development supports that have been found to be effective in improving teachers' practice and/or children's outcomes, we are not able to determine whether changes in children's knowledge and skills are related to 1) improvements in teachers' knowledge, skills, and general classroom practice (e.g., teachers' ability to better perceive and respond to children's cues during moment-to-moment exchanges); 2) teachers' ability to implement the curriculum with a high degree of quality and fidelity; and/or 3) the curriculum or activities themselves.

Planned variation studies and studies testing individual professional development components are evaluation approaches that will help to disentangle the impacts of various professional development components and disaggregate impacts of a curriculum from professional development. Examples of planned variation in approaches can be drawn from language and literacy interventions that have compared approaches to professional development of varying intensity (e.g., Neuman & Wright, 2010; Pianta, Mashburn et al., 2008) or studies that have compared the receipt of a curriculum alone with the receipt of curriculum plus professional development (e.g., Assel, Landry, Swank, & Gunnewig, 2007; Kinzie, Vick Whittaker, Williford et al., 2013). In addition, language and literacy researchers have begun testing less intensive professional development that targets a discrete skill (e.g., Justice et al., 2010), which provides evidence of the link between the professional development and that specific competency.

Investigate the Role of the Teacher as a Learner

In a review of the state of the research on professional development in early childhood programs as a whole, Sheridan and colleagues (2009) called for more research on the personal variables that have an impact on professional development. This is particularly true in the area of early childhood mathematics professional development, where few studies have examined personal characteristics that may mediate or moderate the impact of professional development. We need to know more about what teacher characteristics lead them to respond, or not respond, to specific types of professional development. For example, is a teacher who feels more efficacious in her ability to teach mathematics more or less likely to be responsive to professional development designed to improve mathematics-related instruction? Similarly, we need to know more about what underlying processes in teachers' experience help translate professional development experiences to changes in practice. Research suggests that teachers' beliefs about early childhood mathematics might be a particularly important characteristic to assess and include in evaluations of the impact of professional development.

Investigate the Role of Dosage and Whether and How Professional Programs Can Be Delivered at Scale

Several of the professional development programs described in this chapter included intensive amounts of professional development, including semester-long courses, multiple workshops, and a large number of coaching hours. For example, TRIAD (Clements, Sarama, Spitler et al., 2011) provided 75 hours of out-of-class training in addition to coaching for teachers, and Clements and colleagues (2011) suggest that this amount may be necessary for teachers to implement a high-quality mathematics curriculum. However, this approach may not be scalable given the constraints on teachers' time and associated costs.

Several of the examples provided in this chapter use a technology component as part of their professional development (i.e., Clements, Sarama, Spitler et al., 2011; Kinzie et al., 2012; Sorkin & Preston, 2010). Delivering professional development support via the Internet can help make teacher professional development more scalable and accessible (see Preston, Chapter 6, for further discussion). However, in all the examples described in this chapter, there was also a face-to-face component of the professional development. More research is needed to determine whether a completely web-mediated professional development experience can have an effect on teachers' classroom practice and whether these effects are comparable to what is achieved with blended in-person and online models.

CONCLUSION

Since the National Research Council's 2009 report on mathematics learning in early childhood, there has been a marked increase in professional development aimed at preparing the early childhood work force to engage in interaction and instruction that positively influences children's mathematics knowledge and skills. There have also been recent efforts to evaluate the effectiveness of that professional development on teachers' classroom practices and children's outcomes. However, there remains a lack of understanding about the process through which various types of professional development exert their influence. This is due, in part, to the lack of alignment among professional development inputs; teacher knowledge, skills, and practice; and child outcomes. In this chapter, we outline a model of intentional teaching that could be used to articulate these linkages, which would allow for more precise evaluation of the mechanism through which professional development in early childhood mathematics is leading to change. Progress in the evaluation of early childhood mathematics professional development must also include the development of well-validated measures to assess various components of the model as well as a better understanding of teacher characteristics that contribute to individual differences in the effectiveness of professional development. With continued research in these areas, we are poised to develop an evidence base about the most effective components and processes of early childhood math professional development that can be used to support the early childhood work force in providing high-quality early math experiences for our youngest learners.

FOR REFLECTION AND ACTION

1. In your experience, what are the principal methods of evaluating mathematics-related professional development? It might be useful to document current evaluation practices in your setting and examine these in light of Vick Whittaker's and Hamre's discussion.

2. The authors present a detailed description of a "theory of change," the Intentional Teaching Framework, that describes how professional development may influence teachers' practices. For your work, it may be useful to read more about this specific theory of change or about theories of change in general. One source is this manual from the Annie E. Casey Foundation (http://www.aecf.org/upload/publicationfiles/cc2977k440.pdf); another is Research to Action's extensive resource list (http://www.researchtoaction.org/2011/05/theory-of-change-useful-resources).

3. Vick Whittaker and Hamre describe and reference a number of existing measures that potentially may be used to evaluate changes in teachers' knowledge or practices as a result of professional development. It may be useful for you to look more closely at several of these to see if they might be a good fit with your situation. Most are available online or in the downloadable content provided with this book.

4. Depending on your professional situation, draw on this chapter's insights to develop or revise an evaluation plan designed to determine the effects of math-related professional development. Try out aspects of the evaluation plan and use the results to improve the professional development efforts in which you are involved.

5. The authors make the point that professional development may have different effects on teachers with different kinds of characteristics. You might reflect on your experience working with teachers with diverse education levels and math-related beliefs.

REFERENCES

Anderson, N.A., Barksdale, M.A., & Hite, C.E. (2005). Preservice teachers' observations of cooperating teachers and peers while participating in an early field experience. *Teacher Education Quarterly, 32*(4), 97–117.

Assel, M.A., Landry, S.H., Swank, P.R., & Gunnewig, S. (2007). An evaluation of curriculum, setting, and mentoring on the performance of children enrolled in prekindergarten. *Reading and Writing, 20*(5), 463–494.

Ball, D.L. (1991). Teaching mathematics for understanding: What do teachers need to know about subject matter? In M. Kennedy (Ed.), *Teaching academic subjects to diverse learners* (pp. 63–83). New York, NY: Teachers College Press.

Ball, D.L., Hill, H.C., & Bass, H. (2005). Knowing mathematics for teaching: Who knows mathematics well enough to teach third grade, and how can we decide? *American Educator, 29*(3), 14–46.

Ball, D.L., Thames, M.H., & Phelps, G. (2008). Content knowledge for teaching: What makes it special? *Journal of Teacher Education, 59*(5), 389–407.

Bandura, A. (1986). *Social foundations of thought and action: A social cognitive theory.* Englewood Cliffs, NJ: Prentice-Hall.

Blevins-Knabe, B., Austin, A., Musun, L., Eddy, A., & Jones, R. (2000). Family home care providers' and parents' beliefs and practices concerning mathematics with young children. *Early Child Development and Care, 165*(1), 41–58.

Borko, H., Jacobs, J., Eiteljorg, E., & Pittman, M.E. (2008). Video as a tool for fostering productive discussions in mathematics professional development. *Teaching and Teacher Education, 24*(2), 417–426.

Clements, D.H., & Sarama, J. (2007). Effects of a preschool mathematics curriculum: Summative research on the Building Blocks project. *Journal for Research in Mathematics Education, 38*(2), 136–163.

Clements, D.H., & Sarama, J. (2008). Experimental evaluation of the effects of a research-based preschool mathematics curriculum. *American Educational Research Journal, 45*(2), 443–494.

Clements, D.H., & Sarama, J. (2009). *Learning and teaching early math: The learning trajectories approach.* New York, NY: Routledge.

Clements, D.H., & Sarama, J. (2011). Early childhood mathematics intervention. *Science, 333*(6045), 968–970.

Clements, D.H., Sarama, J., & Liu, X. (2008). Development of a measure of early mathematics achievement using the Rasch model: The research-based early math assessment. *Educational Psychology, 28*(4), 457–482.

Clements, D.H., Sarama, J., Spitler, M.E., Lange, A.A., & Wolfe, C.B. (2011). Mathematics learned by young children in an intervention based on learning trajectories: A large-scale cluster randomized trial. *Journal for Research in Mathematics Education, 42*(2), 127–166.

Clements, D.H., Sarama, J., & Wolfe, C.B. (2011). *TEAM: Tools for early assessment in mathematics.* Columbus, OH: McGraw-Hill Education.

Copple, C.E. (2004). Mathematics curriculum in the early childhood context. In D.H. Clements, J. Sarama, & A.-M. DiBaise (Eds.), *Engaging young children in mathematics: Standards for early childhood mathematics education* (pp. 83–87). Mahwah, NJ: Lawrence Erlbaum Associates.

Dall'Alba G., & Sandberg, J. (2006). Unveiling professional development: A critical review of stage models. *Review of Educational Research, 76*(3), 383–412.

Dewey, J. (1933). *How we think: A restatement of the relation of reflective thinking to the educative process.* Boston, MA: Heath.

Dickinson, D., & Caswell, L. (2007). Building support for language and early literacy in preschool classrooms through in-service professional development: Effects of the Literacy Environment Enrichment Program (LEEP). *Early Childhood Research Quarterly, 22*(2), 243–260.

Ericsson, K.A., & Charness, N. (1994). Expert performance: Its structure and acquisition. *American Psychologist, 49*(8), 725–747.

Ertle, B., Ginsburg, H.P., Cordero, M.I., Curran, T.M., Manlapig, L., & Morgenlander, M. (2008). The essence of early childhood mathematics education and the professional development needed to support it. In A. Dowker (Ed.), *Mathematical difficulties: Psychology and intervention* (pp. 60–84). Amsterdam: Elsevier.

Ertle, B., Ginsburg, H.P., & Lewis, A. (April, 2006). *Measuring the efficacy of Big Math for Little Kids: A look at fidelity of implementation.* Paper presented at the annual meeting of the American Educational Research Association, San Francisco, CA.

Ginsburg, H.P., Kaplan, R.G., Cannon, J., Cordero, M.I., Eisenband, J.G., Galanter, M., & Morgenlanger, M. (2006). Preparing Early Childhood Educators to Teach Mathematics. In M. Zaslow & I. Martinez-Beck (Eds.), *Critical issues in early childhood professional development* (pp. 171–202). Baltimore, MD: Paul H. Brookes Publishing Co.

Ginsburg, H.P., Lee, J.S., & Boyd, J.S. (2008). Mathematics education for young children: What it is and how to promote it. *Society for Research in Child Development Social Policy Report, 22*(1), 1–23.

Hamre, B.K., Downer, J.T., Jamil, F.M., & Pianta, R.C. (2012). Enhancing teachers' intentional use of effective interactions with children. In R.C. Pianta, W.S. Barnett, L.M. Justice, & S.M. Sheridan (Eds.), *Handbook of early childhood education* (pp. 507–532). New York, NY: Guilford Press.

Hamre, B.K., Pianta, R.C., Burchinal, M., Field, S., LoCasale-Crouch, J.L., Downer, J.T., . . . Scott-Little, C. (2012). A course on effective teacher-child interactions: Effects on teacher beliefs, knowledge, and observed practice. *American Educational Research Journal, 49*(1), 88–123.

Hart, L.C. (2002). Preservice teachers' beliefs and practices after participating in an integrated content/method course. *School Science and Mathematics, 102*(1), 4–14.

Hatch, T., & Grossman, P. (2009). Learning to look beyond the boundaries of representation: Using technology to examine teaching (Overview for a digital exhibition: Learning from the Practice of Teaching). *Journal of Teacher Education, 60*(1), 70–85.

Hill, H.C., & Ball, D.L. (2004). Learning mathematics for teaching: Results from California's Mathematics Professional Development Institutes. *Journal for Research in Mathematics Education, 35*(5), 330–351.

Hill, H.C., Blunk, M.L., Charalambous, Y.C., Lewis, J.M., Phelps, G.C., Sleep, L., . . . Ball, D.L. (2008). Mathematical knowledge for teaching and the mathematical quality of instruction: An exploratory study. *Cognition and Instruction, 26*(4), 430–511.

Hill, H.C., Rowan, B., & Ball, D.L. (2005). Effects of teachers' mathematical knowledge for teaching on student achievement. *American Educational Research Journal, 42*(2), 371–406.

Jamil, F.M. (2013, April). *Examining teacher reflection and its association with effective teacher-child interactions.* Poster presented at the biannual Society for Research in Child Development Conference, Seattle, WA.

Jamil, F.M., Sabol, T., Hamre, B.K., & Pianta, R.C. (2012). *Assessing teachers' skill in detecting and identifying effective interactions in the classroom: Theory and measurement.* Manuscript submitted for publication.

Justice, L.M., McGinty, A.S., Cabell, S.Q., Kilday, C.R., Knighton, K., & Huffman, G. (2010). Language and literacy curriculum supplement for preschoolers who are academically at-risk: A feasibility study. *Language, Speech, and Hearing Services in Schools, 41*(2), 161–178.

Kilday, C.R., & Kinzie, M.B. (2009). An analysis of instruments that measure the quality of mathematics teaching in early childhood. *Early Childhood Education Journal, 36*(4), 365–372.

Kinzie, M., Pianta, R.C., Whittaker, J., Foss, M.J., Pan, E., Lee, Y., Williford, A.P., & Thomas, J.B. (2010). MyTeachingPartner Math/Science demonstration video. Charlottesville, VA: MyTeachingPartner Math/Science. http://www.mtpmathscience.net

Kinzie, M.B., Vick Whittaker, J.E., Kilday, C., & Williford, A.P. (2012). Designing effective curricula and teacher professional development for early childhood mathematics and science. In C. Howes, B. Hamre, & R. Pianta (Eds.), *Effective early childhood professional development: Improving teacher practice and child outcomes* (pp. 31–59). Baltimore, MD: Paul H. Brookes Publishing Co.

Kinzie, M.B., Vick Whittaker, J.E., McGuire, P., Lee, Y., & Kilday, C.R. (2013). *Pre-Kindergarten mathematics and science: Design-based research on curricular development.* Manuscript submitted for publication.

Kinzie, M.B., Vick Whittaker, J.E., Williford, A.P., DeCoster, J., McGuire, P., Lee, Y., & Kilday, C.R. (2013). *MyTeachingPartner-Math/Science pre-kindergarten curricula and teacher supports: Associations with children's math and science learning.* Manuscript submitted for publication.

Klein, A., Starkey, P., Sarama, J., Clements, D.H., & Iyer, R. (2008). Effects of a pre-kindergarten mathematics intervention: A randomized experiment. *Journal of Research on Educational Effectiveness, 1*(3), 155–178.

Kolb, D.A., Boyatzis, R.E., & Mainemelis, C. (2000). Experiential learning theory: Previous research and new directions. In R.J. Sternberg & L.E. Zhang (Eds.), *Perspectives on cognitive, learning and thinking styles* (pp. 193–210). Mahwah, NJ: Lawrence Erlbaum Associates.

Kottkamp, R. (1990). Means of facilitating reflection. *Education and Urban Society, 22*(2), 182–203.

Lobman, C., Ryan, S., & McLaughlin, J. (2005). Reconstructing teacher education to prepare qualified preschool teachers: Lessons from New Jersey. *Early Childhood Research & Practice, 7*(7), 1–15.

Martinez-Beck, I., & Zaslow, M. (2006). The context for critical issues in early childhood professional development. In M. Zaslow & I. Martinez-Beck (Eds.), *Critical issues in early childhood professional development* (pp. 171–202). Baltimore, MD: Paul H. Brookes Publishing Co.

Mashburn, A.J., Pianta, R.C., Hamre, B.K., Downer, J.T., Barbarin, O., Bryant, D., . . . Howes, C. (2008). Measures of classroom quality in prekindergarten and children's development of academic, language, and social skills. *Child Development, 79*(3), 732–749.

McCray, J.S., & Chen, J. (2012). Pedagogical content knowledge for preschool mathematics: Construct validity of a new teacher interview. *Journal of Research in Childhood Education, 26*(3), 291–307.

Moreno, R., & Ortegano-Layne, L. (2008). Do classroom exemplars promote the application of principles in teacher education? A comparison of videos, animations, and narratives. *Educational Technology Research & Development, 56*(4), 449–465.

National Association for the Education of Young Children & National Council of Teachers of Mathematics. (2002). *Early childhood mathematics: Promoting good beginnings.* Washington, DC: Author. Retrieved from http://www.naeyc.org/files/naeyc/file/positions/psmath.pdf

National Council for Accreditation of Teacher Education. (2010). *Transforming teacher education through clinical practice: A national strategy to prepare effective teachers.* Washington, DC: Author. Retrieved from http://www.ncate.org/LinkClick.aspx?fileticket=zzeiB1OoqPk%3D&tabid=715

National Mathematics Advisory Panel. (2008). *Foundations for success: The final report of the National Mathematics Advisory Panel.* Washington, DC: U.S. Department of Education. Retrieved from http://www2.ed.gov/about/bdscomm/list/mathpanel/report/final-report.pdf

National Research Council. (2009). *Mathematics learning in early childhood: Paths toward excellence and equity.* Committee on Early Childhood Mathematics, C.T. Cross, T.A. Woods, & H. Schweingruber (Eds.). Center for Education, Division of Behavioral and Social Sciences and Education. Washington, DC: National Academies Press.

Neuman, S.B., & Wright, T. (2010). Promoting language and literacy development for early childhood educators: A mixed-methods study of coursework and coaching. *Elementary School Journal, 111*(1), 63–86.

Newton, S. (2010). *Preservice performance assessment and teacher early career effectiveness: Preliminary findings on the performance assessment for California teachers.* Stanford, CA: Stanford University, Stanford Center for Assessment, Learning, and Equity.

Pajares, M.F. (1992). Teachers' beliefs and educational research: Cleaning up a messy construct. *Review of Educational Research, 62*(1), 307–332.

Pianta, R.C., La Paro, K.M., & Hamre, B.K. (2008). *Classroom Assessment Scoring System™ (CLASS™) Manual, Pre-K.* Baltimore, MD: Paul H. Brookes Publishing Co.

Pianta, R., Mashburn, A., Downer, J., Hamre, B., & Justice, L. (2008). Effects of web-mediated professional development resources on teacher-child interactions in pre-kindergarten classrooms. *Early Childhood Research Quarterly, 23*(4), 431–451.

Preston, M., Campbell, G., Ginsburg, H., Sommer, P., & Moretti, F. (2005). Developing new tools for video analysis and communications to promote critical thinking. In P. Kommers & G. Richards (Eds.), *Proceedings of World Conference on Educational Multimedia, Hypermedia and Telecommunications 2005* (pp. 4357–4364). Chesapeake, VA: Association for the Advancement of Computing in Education.

Raver, C.C., Blair, C., & Li-Grining, C. (2012). Extending models of emotional self-regulation to classroom settings: Implications for professional development. In C. Howes, B.K. Hamre, & R.C. Pianta (Eds.), *Effective early childhood professional development: Improving teacher practice and child outcomes* (pp. 113–130). Baltimore, MD: Paul H. Brookes Publishing Co.

Santagata, R. (2005). Practices and beliefs in mistake-handling activities: A video study of Italian and U.S. mathematics lessons. *Teaching and Teacher Education, 21*(5), 491–508.

Santagata, R., Zannoni, C., & Stigler, J.W. (2007). The role of lesson analysis in preservice teacher education: An empirical investigation of teacher learning from a virtual video-based field experience. *Journal of Mathematics Teacher Education, 10*(2), 123–140.

Sarama, J., & Clements, D.H. (2009). *Manual for classroom observation of early mathematics environment and teaching.* Buffalo: University at Buffalo, State University of New York.

Sarama, J., Clements, D.H., Starkey, P., Klein, A., & Wakeley, A. (2008). Scaling up the implementation of a pre-kindergarten mathematics curriculum: Teaching for understanding with trajectories and technologies. *Journal of Research on Educational Effectiveness, 1*(2), 89–119.

Schank, R. (1982). *Dynamic memory: A theory of reminding and learning in computers and people.* New York, NY: Cambridge University Press.

Sheridan, S., Edwards, C., Marvin, C., & Knoche, L. (2009). Professional development in early childhood programs: Process issues and research needs. *Early Education and Development, 20*(3), 377–401.

Shulman, L.S. (1987). Knowledge and teaching: Foundations of the new reform. *Harvard Educational Review, 57*(1), 1–22.

Sorkin, J.E., & Preston, M.D. (2010, May). *Designing a video library and a web environment for learning about early childhood mathematics education.* Paper presented at the American Educational Research Association, Denver, CO.

Star, J.R., & Strickland, S.K. (2008). Learning to observe: Using video to improve preservice mathematics teachers' ability to notice. *Journal of Mathematics Teacher Education, 11*(2), 107–125.

Starkey, P., Klein, A., & Wakeley, A. (2004). Enhancing young children's mathematical Knowledge through a pre-kindergarten mathematics intervention. *Early Childhood Research Quarterly, 19*(1), 99–120.

Van Es, E.A., & Sherin, M.G. (2002). Learning to notice: Scaffolding new teachers' interpretations of classroom interactions. *Journal of Technology and Teacher Education, 10*(4), 571–596.

Vick Whittaker, J.E., Kinzie, M.B., Williford, A.P., & DeCoster, J. (2013, April). *Trajectories of math and science teaching quality: Effects of MyTeachingPartner-Math/Science.* Paper presented at the biannual Society for Research in Child Development Conference, Seattle, WA.

Wilkins, J.L.M., & Brand, B.R. (2004). Change in preservice teachers' beliefs: An evaluation of a mathematics methods course. *School Science and Mathematics, 104*(5), 226–232.

Yesil-Dagli, U., Lake, V.E., & Jones, I. (2010). Preservice teachers' beliefs about mathematics and science content and teaching. *Journal of Research in Education, 20*(2), 32–48.

Zaslow, M., Tout, K., Halle, T., Whittaker, J.E., & Lavelle, B. (2010). Emerging research on early childhood professional development. In S.B. Neuman & M.L. Kamil (Eds.), *Preparing teachers for the early childhood classroom: Proven models and key principles* (pp. 19–47). Baltimore, MD: Paul H. Brookes Publishing Co.

9

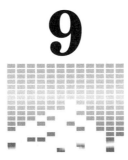

The Future?

Herbert P. Ginsburg, Taniesha A. Woods, and Marilou Hyson

This chapter describes a road map to the future. The goals are clear: We need to help teachers teach math more effectively than they do now so that children can learn it more deeply, extensively, and even happily than they do now. Achieving these goals will not be simple or easy, but fortunately, we can draw on valuable resources and models of professional development to begin the journey toward quality early childhood mathematics education (ECME) for all. Here is an itinerary.

RECRUIT TALENTED INDIVIDUALS TO THE PROFESSION

To start, we need to recruit talented women and men to participate in early childhood education, a profession that should be seen (but is not now) as prestigious and honorable, with compensation and benefits proportionate to its importance. Even more than in other areas of education, the early childhood work force has historically been dominated by women, to a great extent because their access to other professions was limited and because teaching young children was perceived as "women's work." Today, with more options open, the strongest women students often choose more lucrative, higher-status careers in law, medicine, or business over early childhood education. It is wonderful that women in particular can now freely enter those higher-status professions. But an unfortunate and unintended consequence is that fewer talented women than in earlier periods of our history choose to become teachers, especially in early childhood education. Hence we need to undertake major efforts to recruit and retain a diverse group of talented, committed individuals—women and men—to teach the young and serve as role models for children. One obvious part of the solution is adequate pay and health care. However, Kagan and

Gomez (Chapter 1) show that current conditions often fall far short of the mark, undermining our efforts before the professional development journey can even begin.

SUPPORT AND RESPECT TEACHERS

Teachers and other early care and education providers who devote themselves to the well-being and education of children should be respected and supported in their professional work. The support must include not only adequate preservice and in-service professional development, as we argue here, but also reasonable class size and working conditions, a sufficient number of aides and other support staff, and comfortable physical environments with a rich array of materials to support learning in mathematics and other domains. Just as physicians need the best tools to do their best work, so do teachers of young children. We need to do much better for a profession vital to the health of our society.

USE HIGHER EDUCATION COURSES AND FIELD EXPERIENCES TO BUILD CORE EARLY CHILDHOOD MATHEMATICS EDUCATION CONCEPTS AND SKILLS

Although it is a lifelong process, professional development in ECME should begin within quality higher education programs. Students in associate and baccalaureate programs need to learn many important ideas and skills—reflected in the "goals for teachers" presented by Juanita Copley in Chapter 4. The five following areas should receive special attention in the future.

Students Must Understand the Development of Children's Mathematical Thinking and Learning

The prospective teacher's journey (or the journey of practicing teachers who have now enrolled in higher education, such as many Head Start staff) should begin with a deep understanding of children's mathematical thinking and how it develops. As Herbert P. Ginsburg's stories about Ben vividly illustrated (Chapter 3), young children have complex mathematical minds that are both abstract and concrete. Children can and do learn a meaningful everyday mathematics that involves not only memory but also important skills and concepts. No doubt, further research on this topic (as on almost all others) is necessary. But cognitive psychology is already powerful enough to serve as a foundation for early math education. We already know a lot and should use what we know.

At the same time, cognitive psychology alone is not sufficient. Recently, for example, several studies have shown that "executive function"—the ability to handle cognitive complexity, to plan, and to control impulses—may play an important role in children's math learning (e.g., Blair & Razza, 2007; Raver et al., 2011). The implication of this research, which is an area ripe for further investigation, is that professional development should not abandon study of the "whole child" in favor of the narrowly "cognitive child."

Future teachers' appreciation of the complexity of children's mathematical thinking can provide two benefits. One is that teachers will understand why ECME is developmentally appropriate and why helping children learn math can satisfy their curiosity and stimulate thought. Teachers need to learn that part of the whole child is the mathematics learner. The other benefit is that teachers will be able to understand individual children's processes of mathematics learning. The cognitive psychology of mathematical thinking provides a theory-based specificity that allows teachers to conduct formative assessments to uncover the reasons for a particular child's response under a given set of conditions. The resulting understanding of the child's thinking can then guide appropriate instruction. The motto for both professional development providers and teachers should be to avoid vague slogans and understand the individual.[1]

Students Must Understand Early Childhood Mathematics Education Content

Higher education students need to learn the content of ECME on two related levels. One involves learning that the math curriculum for young children should cover important mathematical skills and ideas—not simply the learning of basic counting and the names of a few shapes. The Common Core State Standards (National Governors Association Center for Best Practices & Council of Chief State School Officers, 2010) can provide guidance on *what* should be taught to children from kindergarten onward (although they do not and should not mandate *how* it should be taught). We believe that the Common Core articulates useful aspirational goals with implications for teaching children younger than kindergarten age as well: Our children should learn much more and deeper math than they do now.[2] To make these aspirations a reality, a second level of learning ECME content is required: Prospective teachers need to know the math to be taught. It is more complex than generally assumed, and many teachers know it only superficially.

Students Must Understand the Structure of Early Childhood Mathematics Education

Higher education courses should help students understand the structure and the organization of ECME. They should understand that ECME involves many elements, including free play, everyday activities in which teachers exploit inherent math content, organized projects, and mathematics curriculum. They need to learn that all these elements can be useful and that planned curriculum in early childhood mathematics can be both developmentally appropriate and extremely productive. At the same time, the setting of higher education is not the place to learn about only one particular curriculum to the exclusion of all others. Rather, future teachers should be given opportunities to explore and evaluate the characteristics and strengths of several curriculum approaches.

Students Must Understand and Practice Intentional Teaching

Students need to learn that intentional teaching is necessary and can be developmentally appropriate (Copple & Bredekamp, 2009; Epstein, 2014). Intentional teaching does not necessarily imply mindless drill or rigid direct instruction. Skilled intentional teaching can promote learning, conceptual understanding, and development of strategies and skills. Educators should certainly support free play and other activities as pathways to mathematical competence, but teachers do not fulfill their responsibilities unless they also engage in sound intentional teaching.

Students Must Learn What Assessment Is and How to Do It

Formative assessment—finding out what a child knows in order to teach her effectively—is at the heart of educational practice. Yet this kind of assessment is seldom given adequate emphasis in professional development, a gap emphasized by Martha Zaslow's (Chapter 5) review of research on the features of effective professional development. How can you teach well if you do not know what the child knows and does not know? In light of a careful assessment of a child's difficulty with a specific mathematical idea, the teacher may try one approach, assess again, and perhaps then try another until the results of assessment show success. Higher education courses should teach students how to observe, conduct clinical interviews, and test children's mathematical competence. Each of these methods has a legitimate place in ECME. Students should also learn about the benefits and dangers of specific methods for evaluating progress, such as achievement testing and high-stakes assessment of young children. Accountability and evaluation are necessary but can be, and often are, counterproductive and pernicious. There may not be much that individual teachers can do about this, but they should understand the issues and consider their role as informed advocates for young children's well-being.

HIGHER EDUCATION'S POOR TRACK RECORD IN PREPARING EARLY CHILDHOOD MATHEMATICS TEACHERS

As we have seen, there is much for future teachers to learn in their higher education programs. Yet, as Marilou Hyson and Taniesha A. Woods (Chapter 2) describe, high-quality college courses dedicated to ECME are rare. Some teacher educational programs offer no such courses; others offer courses that barely cover ECME, ignore key aspects of the content, or lack opportunities for the application of knowledge. We desperately need new or radically redesigned courses and related field experiences designed to help teachers in training to gain in-depth knowledge and skills in the five areas just described, from the cognitive psychology of mathematical thinking to various methods and uses of assessment. Mathematics education courses should also engage students in multiple field experiences under the supervision of master teachers or coaches.

THE ROAD TO INNOVATION IN PRESERVICE COURSES AND FIELD EXPERIENCES

Fortunately, Michael D. Preston's chapter (Chapter 6) on preservice professional development provides guidelines for developing evidence-based, innovative courses and field experiences. Using online illustrations, he shares examples of successful or promising pedagogy, many of which involve innovative uses of technology. As Preston describes and illustrates, ECME courses can offer opportunities for students to analyze videos showing children's mathematical thinking and the processes of intentional teaching that help children learn skills and concepts while building their interest and engagement.

Besides watching others teach, the courses we need should also provide opportunities for students to participate in experiences that connect, in Martha Zaslow's terms, "knowledge-based" and "practice-based" learning. As emphasized in almost every chapter in this book, there is ample evidence that teachers cannot become effective practitioners in math or in any other domain without these kinds of rich, well-supervised, on-the-ground experiences. To accomplish this, teamwork will be needed: Professors who teach ECME courses should play a role in field supervision, and early childhood teachers, math coordinators, and the like should play a role in college courses. The overall goal should be to help the student bridge the gap between higher education and the classroom, and the chasm between theory and practice, so that the college experience becomes less "academic," in the sense of an abstract exercise with little practical value—a frequent criticism of teacher education in general, not only of teacher education in early childhood mathematics.

Rich and meaningful ECME courses and related field experiences are possible, but only if professors are prepared to implement them properly. As described by Hyson and Woods (Chapter 2), many faculty members do not know the ECME material to be taught or the most effective pedagogy for teaching this material. After all, many of them have never taken a course of this type either. With this gap in mind, we need to develop ways to help upgrade the skills of early childhood faculty who are or will be involved in ECME. Special workshops or summer institutes may be effective, but online faculty resources may produce a broader impact at substantially lower cost. Greatly increased investments in faculty development, including evaluation of these initiatives, will be necessary, but the effort will be worthwhile because the impact on faculty competence, graduates' skills, and therefore children's outcomes is likely to be substantial.

Develop Extensive and Intensive Systems of In-Service Professional Development

Although we wish for an early childhood work force in which all teachers have had high-quality college courses and field experiences, for the foreseeable future the reality is that many will lack that foundation. Whatever their prior educational background, practicing teachers and other early care and education providers need robust, ongoing support for their work. They need to perfect, or in many cases learn for the first time, what college courses should have taught them—namely, an understanding of the development of children's mathematical thinking and learning, the math itself, the content and structure of ECME, and the nature of intentional teaching and assessment. But even graduates of the best higher education programs have much to learn on the job: They must become increasingly skilled and confident in the daily implementation of intentional teaching and the formative assessment of children's mathematical competence. And most of them must now do something that their college program did not, and probably should not, prepare them to do: implement one specific mathematics curriculum. As noted previously, preservice students' courses should help them become familiar with a variety of curriculum approaches. Later, on-the-job training is needed in order to implement a math curriculum specific to the setting in which they work.

Fortunately, we know a great deal about how to help teachers learn these things after they leave higher education. As Kimberly Brenneman shows in Chapter 7, in-service professional development should be extensive and intensive. Teachers should meet as a community as often as they can to discuss their own teaching of early childhood mathematics. They should receive regular support from coaches, with sufficient dosage to make a difference in their practice. As in preservice courses, technology can also help: Practicing teachers should learn to analyze videos of themselves teaching and of their formative assessment of children in their classrooms. And as Brenneman describes, they should learn to pay special attention to the needs of children learning English as a second language. The innovative, well-evaluated approaches Brenneman depicts should be essential guides for our road map toward the future of in-service professional development, as are her sensible cautions about possible obstacles to their implementation.

In-service professional development serves a wide range of practitioners, working in varied settings with equally varied prior knowledge. Fortunately, those who plan and implement in-service professional development in the future can also make use of a hybrid collection of resources, some of which Brenneman describes and others that are referred to throughout various chapters in this book. For example, those teachers who are working on an advanced degree can take ECME college courses that are increasingly becoming available online. Other practicing teachers can receive such instruction through ECME workshops, coaching sessions, and other continuing education work—with the caution that these experiences should be consistent with the research on effective professional development. Web-based material such as that referred to in our discussion of preservice professional development can also be useful for practicing teachers. Like their colleagues enrolled in higher education programs, there is no reason they could not benefit from web-based courses, modules, demonstration videos, clinical interview lessons, and the like. As seen in Brenneman's summary, evidence from several new professional development initiatives in early childhood math and science shows us the potential for the future.

Repeatedly, this book's chapters have emphasized that in-service professional development must be intensive to be effective. In planning for the future, adding a workshop or two during the academic year will be of little value. Intensive, consistent, and skillful support is needed. This, of course, takes time and costs money, but good things are not cheap—and with the use of new technologies some of the cost can be contained.

When thinking about potential participants in in-service professional development, it is worth noting that educational policy makers, administrators, program directors, principals, and specialists in other areas can also benefit from understanding children's learning of mathematics. Their needs are different, and therefore, the professional development content and methods will need to differ from what is provided to practicing early childhood teachers. But increased knowledge of children's mathematical development and of effective approaches to teaching mathematics will increase the ability of this larger group of educators to support and advocate for improved ECME. The authors represented here have aimed at a wide audience; this audience may benefit not only from this book's content but also from specialized institutes, workshops, and online resources intended to build their engagement in ECME.

Conduct and Improve Sound Evaluations of Professional Development Efforts

We owe it to the children we serve (and to those who pay for the services we provide) to implement ECME that is effective for all young children. Part of our responsibility, then, is to evaluate our professional development programs—both preservice and in-service—so that we can improve them and ultimately improve children's achievement. Without the guidance obtained from such evaluations, we may well end up driving rapidly in the wrong direction, guided by a faulty road map to the future. Zaslow (Chapter 5) and Jessica Vick Whittaker and Bridget K. Hamre (Chapter 8) note that in the past, such evaluations of early childhood professional development have been rare. Conducting rigorous and informative evaluations is not easy. The validity and usefulness of evaluations depend on several factors.

First, evaluation of the effectiveness of ECME professional development requires sound measures of teacher knowledge, attitudes, beliefs, and most important, skill in intentional teaching. This, in itself, is a huge undertaking. In Chapter 8, Vick Whittaker and Hamre describe groundbreaking efforts to deal with the measurement of these important factors, efforts that should inform future evaluation practices.

Second, evaluation requires understanding and measurement of the different components of professional development. It is not enough to say that the professional development "works" in improving teaching practice. That is important to know, but it is also important to know what professional development components improve teaching and how they accomplish the job. Only then can we improve specific professional development practices and components. Yet most of the research in this area has not gone beyond the evaluation of the effects of loosely specified professional development programs on equally loosely specified outcomes. This is not a criticism: Research in this area is new and the research challenges are immense.

Third, evaluation of professional development's effectiveness requires sound measures of children's learning. The ultimate criterion of effectiveness is whether professional development helps teachers enough so that they can improve young children's knowledge of and attitudes toward math.

Despite these worthy efforts, a great deal needs to be done to develop sound evaluations. This effort is in its infancy, especially in early childhood mathematics. Considerable development and research are needed to create useful measures of teacher knowledge and practice, and of children's learning. And theoretical work and research are required to understand the components and processes involved in professional development.

One final caution: Under certain conditions, evaluation—whether of professional development or curricula—can have a destructive effect on ECME. If we evaluate professional development using inadequate measures of teaching and learning, teachers will eventually learn to game the teaching measures and will also teach to the overly narrow or developmentally inappropriate child outcome measures. The danger is that the methods of evaluation may distort and degrade teaching and learning—another example of an inaccurate road map leading to the wrong destination.

PUBLIC ENGAGEMENT

The Introduction to this book focused on professional development as a tool to prepare teachers to give young children good math experiences. But young children are only in early childhood programs for part of the day, if at all. For ECME to succeed, *everyone* needs to understand and get on board with its goals and methods: not only professional educators but also families, health care providers, community leaders, politicians, and the public at large. Just as there have been major efforts at public engagement in promoting early literacy development, so should similar efforts be employed in early mathematical development.

When engaging families, it is especially important to partner with those who have often been alienated from education, many because their own educational experiences were miserable. Many families and others in the community have fixed, usually negative ideas about mathematics. In turn, these influence what families may do at home and what is supported in public policy. For example, if you think that the primary goal of mathematics education is to drill young children in facts and skills, you may favor investing in shoot-down-the-number-facts-from-the-sky software that is little more than an electronic drill sheet. But if you are helped to understand and value children's mathematical thinking, you will want to find out ways to engage children in very different kinds of meaningful activities. Families, communities, and the public at large need opportunities to be empowered in supporting young children's math learning.

How can these different constituencies learn about and become engaged in ECME? Families can be reached through the early childhood programs that their children attend, using proven strategies to enhance family engagement. Parent, family, and community engagement has long been a cornerstone of Head Start and a major contributor to its effectiveness. Many early childhood math curricula offer links to families—for example, take-home letters about what children are learning and activities to do at home. Technology also serves as a powerful tool that families can use to engage their children in ECME. Local Public Broadcasting Service (PBS) stations, for example, often have television programming, web sites, and mobile apps for cell phones and tablets that families and children use; these platforms could offer materials on math as well as on other topics of interest to families. The Family Math project conducts workshops for parents and offers written materials that include math activities to do at home (Coates & Stenmark, 1997; Coates & Thompson, 1999). Health care providers are an untapped resource in ECME; in the domain of literacy, for example, the Reach Out and Read program involved pediatricians by giving them resource materials for the waiting room and storybooks for them to send home with families after well-child checkups. Why not something similar in math?

Policy makers may have received greater background knowledge in literacy than in mathematics, but with the recent emphasis on science, technology, engineering, and math, that is likely to change. Concise, evidence-based briefing papers, visits to members of state and federal legislative bodies, and compelling anecdotes about the value of engaging all young children in mathematics are potentially effective ways to reach this key group of stakeholders.

A major challenge is the engagement of the public at large. Most people have negative memories of math in their own childhood, probably more so than their memories of books and reading. Building on earlier discussions of the power of video, perhaps public service announcements and YouTube clips can entice even math-averse citizens into smiling and laughing at young children's joy in their mathematical explorations.

THE POLITICAL PROCESS

It is one thing to understand the goals and methods of ECME; it is quite another to value them. Also important is the willingness to pay for what is needed. The very visible elephant in the room, acknowledged at various points in this book, is that implementing an effective program of ECME and building the necessary, extensive supports for prospective and practicing teachers will cost a lot of money. But surely the education of our children should be one of our highest priorities.

Decisions concerning priorities and investments in ECME, including priorities and investments in professional development, play out through the political process, in state legislatures, in Congress, and in the executive branch. Taking the long-term view, there are reasons for optimism. One is that public policy has begun to shift. Several professional organizations and government agencies—for example, the Office of Head Start in the U.S. Department of Health and Human Services—have advocated for new views of ECME and have reflected this enhanced emphasis in new teacher qualification requirements, curriculum standards, evaluation mandates, and professional development expectations. A related reason for optimism is that the Obama administration has proposed a major initiative, in partnership with states and supported by impressive levels of funding, to provide universal high-quality preschool education. We argue that preschool programs cannot be considered of high quality unless they include sound mathematics education.

CONCLUSION

The transformation of professional development in ECME that is described and advocated in this book cannot occur in a vacuum. With our fellow authors, we hope that our vision for the future will become reality, guided by the research and innovative practices that we have described, and supported by broad civic engagement and sound public policies.

FOR REFLECTION AND ACTION

1. Chapter 9, written by the editors, provides a conclusion to this book. This might be a good time to look back through the preceding chapters. From your professional perspective,

which chapters and which sections of chapters have given you the most useful insights for your future work? Which have given you new ideas or specific resources to follow up, on your own or perhaps with colleagues?

2. The chapter offers a road map or itinerary toward the future of professional development in early childhood mathematics. Out of necessity, the road map is large and general. Thinking about your organization or work setting, what might your road map look like?

3. The authors emphasize that a first step in improving professional development is to improve the recruitment and retention of talented, dedicated early childhood educators. To what extent is that necessary in your environment? And if so, to what extent is it feasible (e.g., in financial terms)? What steps might be taken in the future to address the issues proactively?

4. The chapter argues that concerted action will be needed by multiple stakeholders if more effective professional development is to occur. Your role may be that of a professional development provider (pre- or in-service), a district or regional coordinator or other administrator, a state or federal policy maker, or any one of a number of other professional positions. Within that position, what specific actions can and will you take to ensure that all young children gain competence and interest in mathematics because of the efforts of well-prepared teachers?

5. Finally, in their road map to the future, the authors envision a much broader commitment in ECME, going beyond classroom walls to include engagement by parents, health care providers, community leaders, and others with a stake in children's futures. Again, within your community, how might such engagement be promoted, and what could your role be?

CHAPTER 9 NOTES

1. As Nabokov wrote, "In art as in science there is no delight without the detail . . . All 'general ideas' (so easily acquired, so profitably resold) must necessarily remain but worn passports allowing their bearers shortcuts from one area of ignorance to another" (Remnick, 2006, p. 244).
2. In our view, the Common Core should be seen as primarily aspirational. It sets useful content goals. The questions of how to teach the content and evaluate students' mastery of it raise separate issues. Implementing the Common Core and assessing student achievement need not involve narrow instruction or mindless testing, both of which are destructive for ECME, or indeed education at any level.

REFERENCES

Blair, C., & Razza, R.P. (2007). Relating effortful control, executive function, and false belief understanding to emerging math and literacy ability in kindergarten. *Child Development, 78*(2), 647–663.

Coates, G.D., & Stenmark, J.K. (1997). *Family math for young children: Comparing.* Berkeley, CA: EQUALS Publications.

Coates, G.D., & Thompson, V. (1999). Involving parents of four- and five-year-olds in their children's mathematics education: The FAMILY MATH experience. In J.V. Copley (Ed.), *Mathematics in the early years* (pp. 205–214). Reston, VA: National Council of Teachers of Mathematics.

Copple, C., & Bredekamp, S. (Eds.). (2009). *Developmentally appropriate practice in early childhood programs serving children from birth through age 8* (3rd ed.). Washington, DC: National Association for the Education of Young Children.

Epstein, A.S. (2014). *The intentional teacher: Choosing the best strategies for young children's learning.* Rev. ed. Washington, DC: National Association for the Education of Young Children.

National Governors Association Center for Best Practices & Council of Chief State School Officers. (2010). *Common Core State Standards.* Washington, DC: Authors.

Raver, C.C., Jones, S.M., Li-Grining, C., Zhai, F., Bub, K., & Pressler, E. (2011). CSRP's impact on low-income preschoolers' preacademic skills: Self-regulation as a mediating mechanism. *Child Development, 82*(1), 362–378.

Remnick, D. (2006). *Reporting: Writings from the New Yorker.* New York, NY: Alfred A. Knopf.

Appendix

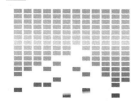

Syllabus for HUDK 4027—Fall 2013

The Development of Mathematical Thinking

Herbert P. Ginsburg

The course to which this syllabus refers is a hybrid designed for both psychology and education students, mostly at the master's-degree level, but also doctoral. The emphasis is heavily psychological, because I believe that understanding the child is the key to math education. Yet I recognize that the course could benefit from much more material on teaching and curriculum. So please take this syllabus as a limited and imperfect example of what a course on early mathematics education should entail. At the same time, a high-quality course in this area should include a good dose of the kind of psychological material emphasized here. I hope that elements of this syllabus can inform the development of effective early mathematics education courses. One key feature of the class that is not described in the syllabus is the extensive use of video, as discussed in Chapter 3 of this book.

1. SEPTEMBER 4: INTRODUCTION AND BACKGROUND

Readings are presented in the order in which they should be read:
Balfanz, R. (1999). Why do we teach children so little mathematics? Some historical considerations. In J.V. Copley (Ed.), *Mathematics in the early years* (pp. 3–10). Reston, VA: National Council of Teachers of Mathematics.
Read this in detail. It's a really interesting history.
Ginsburg, H.P. (2013). Introduction. In H.P. Ginsburg (Ed.), *Children's math*. Manuscript in preparation.
National Research Council. (2009). *Mathematics learning in early childhood: Paths toward excellence and equity*. Committee on Early Childhood Mathematics, C.T. Cross,

---

From Ginsburg, H. (2013). *Syllabus for HUDK 4027—Fall 2013: The development of mathematical thinking*. New York, NY: Columbia University; reprinted by permission.

T.A. Woods, & H. Schweingruber (Eds.). Center for Education, Division of Behavioral and Social Sciences and Education. Washington, DC: National Academies Press.

This is the latest National Academy of Science report on early childhood math education. Although the entire report is worth reading, for now, just read the executive summary (pp. 14–16). You can download the entire book for free at the National Academy Press (http://books.nap.edu/catalog.php?record_id=12519).

2. SEPTEMBER 11: EVERYDAY MATHEMATICS

Ginsburg, H.P. (2013). Chapter 2: Everyday math. In H.P. Ginsburg (Ed.), *Children's math*. Manuscript in preparation.

Nunes, T., Schliemann, A.D., & Carraher, D.W. (1993). Arithmetic in the streets and in schools. In T. Nunes, A.D. Schliemann, & D.W. Carraher (Eds.), *Street mathematics and school mathematics* (pp. 13–27). Cambridge, England: Cambridge University Press.

This is an account of some fascinating cross-cultural work on a pragmatic, everyday mathematics that is very different from the type usually seen in play.

Kontos, S. (1999). Preschool teachers' talk, roles, and activity settings during free play. *Early Childhood Research Quarterly, 14*(3), 363–382.

This examines what teachers do to build on or seize the teachable moments in children's free play; the answer is very little.

3. SEPTEMBER 18: EARLY NUMBER: CONCEPTS, COUNTING, AND HOW MANY

Talbot, M. (2006, September 4). The Baby Lab. *New Yorker, 82*(27), 90–101.

This is an example of good science writing that describes the work of Elizabeth Spelke, a prominent researcher on early number. The main issue to think about is the exact meaning of successful performance on the various tasks she uses to test "knowledge" of mathematics.

Ginsburg, H.P. (2013). Chapter 3: Concepts, counting, and how many. In H.P. Ginsburg (Ed.), *Children's math*. Manuscript in preparation.

Gelman, R. (2000). The epigenesis of mathematical thinking. *Journal of Applied Developmental Psychology, 21*(1), 27–37.

Some of this material is a bit difficult, but it will introduce you to the thinking of Rochel Gelman, who is one of the major researchers in the area of early mathematics.

Alibali, M.W., & DiRusso, A.A. (1999). The function of gesture in learning to count: More than keeping track. *Cognitive Development, 14*(1), 37–56.

Embodied cognition sheds new light on key cognitive issues.

4. SEPTEMBER 25: EARLY NUMBER: ADDING AND TAKING AWAY

Ginsburg, H.P. (2013). Chapter 4: Adding and taking away. In H.P. Ginsburg (Ed.), *Children's math*. Manuscript in preparation.

National Research Council. (2009). *Mathematics learning in early childhood: Paths toward excellence and equity*. Committee on Early Childhood Mathematics, C.T. Cross, T.A. Woods, & H. Schweingruber (Eds.). Center for Education, Division of Behavioral and Social Sciences and Education. Washington, DC: National Academies Press.

Read pages 5-11 to 5-42. This is an overview of how addition grows out of the child's early knowledge. It focuses on the development of relations and operations, particularly in children from age 4 to Grade K. We review issues of symbolism later. Also see the boxes at the end, particularly Boxes 5-2, 5-10, and 5-11.

Zur, O., & Gelman, R. (2004). Young children can add and subtract by predicting and checking. *Early Childhood Research Quarterly, 19*(1), 121–137.

This is an example of how Gelman's work can be applied to a classroom setting. What do you think of the use of prediction as a central task?

5. OCTOBER 2: CLINICAL INTERVIEW

Piaget, J. (1967). Problems and methods. In J. Piaget (Ed.), *The child's conception of the world* (pp. 1–32). Totowa, NJ: Littlefield, Adams.

This is a classic work on the subject—the only thing Piaget ever wrote about clinical interview, as far as I know.

Ginsburg, H.P. (1997). *Entering the child's mind: The clinical interview in psychological research and practice*. New York, NY: Cambridge University Press.

Read the following two chapters: "What is the clinical interview?" (chapter 2) and "Not a cookbook: Guidelines for conducting a clinical interview" (chapter 4). The issue now is how you are going to conduct an interview. These chapters are intended to help. Good luck. Piaget said you need a year of daily practice!

Ellemor Collins, D.L., & Wright, R.J. (2008). Assessing student thinking about arithmetic: Videotaped interviews. *Teaching Children Mathematics, 15*(2), 106–111.

This is an example of using the interview in the classroom. And now, even a smartphone is a videocamera.

Goldin-Meadow, S. Beyond words: The importance of gesture to researchers and learners. *Child Development, 2000*(71), 231–139.

Keep your eye on their hands (and yours too)!

6. OCTOBER 9: SYMBOLS

Ginsburg, H.P. (2013). Chapter 5: Symbolism. In H.P. Ginsburg (Ed.), *Children's math*. Manuscript in preparation.

Miller, K.F., & Parades, D.R. (1996). On the shoulders of giants: Cultural tools and mathematical development. In R.J. Sternberg & T. Ben-Zeev (Eds.), *The nature of mathematical thinking* (pp. 83–117). Mahwah, NJ: Lawrence Erlbaum Associates.

This book focuses on the role of culture in the development of symbolism. The selected chapter is particularly strong on the role of Asian symbol systems. Use this reading to think about why we have symbols and what influence culture may have on them.

Seo, K.-H., & Ginsburg, H.P. (2003). "You've got to carefully read the math sentence . . .": Classroom context and children's interpretations of the equals sign. In A.J. Baroody & A. Dowker (Eds.), *The development of arithmetic concepts and skills: Recent research and theory* (pp. 161–187). Mahwah, NJ: Lawrence Erlbaum Associates.

This is one of the few studies that examines the triangle of children's minds, teacher practice, and classroom materials.

7. OCTOBER 16: NUMBER FACTS AND WRITTEN CALCULATION

Clinical Interview Assignment Due

Ginsburg, H.P. (2013). Chapter 6: Number facts and calculation. In H.P. Ginsburg (Ed.), *Children's math*. Manuscript in preparation.

This is an account of algorithms, invented strategies, and nasty little bugs.

Baroody, A.J., Bajwa, N.P., & Eiland, M. (2009). Why can't Johnny remember the basic facts? *Developmental Disabilities Research Reviews, 15*(1), 69–79.

This is a good overview of the complexity of learning number facts.

Locuniak, M.N., & Jordan, N.C. (2008). Using kindergarten number sense to predict calculation fluency in second grade. *Journal of Learning Disabilities, 41*(5), 451–459.

This article discusses the application of ideas about number sense to learning difficulties.

Trivett, J. (1980). The multiplication table: To be memorized or mastered? *For the Learning of Mathematics, 1*(1), 21–25.

Tables are wonderful ways to teach number facts. This gets a little mathematically complicated, but it is a good example.

Sun, W., & Zhang, J. (2001). Teaching addition and subtraction facts: A Chinese perspective. *Teaching Children Mathematics, 8*(1), 28–31.
This is said to be a Chinese approach and very different from Math Blasters.

8. OCTOBER 23: THE MANY FACES OF UNDERSTANDING

Ginsburg, H.P. (2013). Chapter 7: Understanding. In H.P. Ginsburg (Ed.), *Children's math*. Manuscript in preparation.
This is an account of what understanding involves.
Schoenfeld, A. (1987). What's all the fuss about metacognition? In A. Schoenfeld (Ed.), *Cognitive science and mathematics education* (pp. 189–215). Hillsdale, NJ: Lawrence Erlbaum Associates.
A classic piece on mathematical metacognition.
Fuson, K.C., Kalchman, M., & Bransford, J.D. (2005). Mathematical understanding: An introduction. In M.S. Donovan & J.D. Bransford (Eds.), *How students learn: Mathematics in the classroom* (pp. 217–256). Washington, DC: National Academies Press.
Written for teachers, this is a balanced introduction to the issue of understanding.
Jordan, N.C., Kaplan, D., Nabors Olah, L., & Locuniak, M. (2006). Number sense growth in kindergarten: A longitudinal investigation of children at risk for mathematics difficulties. *Child Development, 77*(1), 153–175.
How is this paper relevant to the issue of understanding?

9. OCTOBER 30: SHAPE AND SPACE

Clements, D.H. (1999). Geometric and spatial thinking in young children. In Copley, J.V. (Ed.), *Mathematics in the early years* (pp. 63–87). Reston, VA: National Council of Teachers of Mathematics.
A basic account of young children's approach to shape and space.
Lehrer, R., Jacobson, C., Kemeny, V., & Strom, D. (1999). Building on children's intuitions to develop mathematical understanding of space. In E. Fennema & T.A. Romberg (Eds.), *Mathematics classrooms that promote understanding* (pp. 63–87). Mahwah, NJ: Lawrence Erlbaum Associates.
This constructivist approach to the teaching of space is based on the idea that instruction should build on what children already know and do not know.
Teppo, A. (1991, March). Van Hiele levels of geometric thought revisited. *Mathematics Teacher,* 210–221.
This is a classic account of the broad development of geometric thinking that also applies to old people (i.e., above second grade).
Ehrlich, S.B., Levine, S., & Goldin-Meadow, S. (2006). The importance of gesture in children's spatial reasoning. *Developmental Psychology, 42*(6), 1259–1268.
This provides another link between math concepts and gesture.

Recommended

National Research Council. (2009). *Mathematics learning in early childhood: Paths toward excellence and equity* (pp. 182–195). Committee on Early Childhood Mathematics, C.T. Cross, T.A. Woods, & H. Schweingruber (Eds.). Center for Education, Division of Behavioral and Social Sciences and Education. Washington, DC: National Academies Press.

10. NOVEMBER 6: PATTERNS AND ALGEBRA

(No more comments on individual papers.
By now, you know that they are all fascinating.)

Blanton, M.L., & Kaput, J.J. (2003, October). Developing elementary teachers' "Algebra eyes and ears." *Teaching Children Mathematics, 10*(2), 70–77.

Warren, E., & Cooper, T.J. (2008). Patterns that support early algebraic thinking in the elementary school. In C.E. Greenes & R. Rubenstein (Eds.), *Algebra and algebraic thinking in school mathematics: Seventieth yearbook* (pp. 113–126). Reston, VA: National Council of Teachers of Mathematics.

Economo-Poulos, K. (1998, December). What comes next? The mathematics of pattern in kindergarten. *Teaching Children Mathematics, 5*(4), 230–233.

Ferrini-Mundy, J., Lappan, G., & Phillips, E. (1997, December). Experiences with patterning. *Teaching Children Mathematics, 3*, 282–288.

Taylor-Cox, J. (2003, January). Algebra in the early years? Yes! *Young Children*, 14–21.

11. NOVEMBER 13: TEACHING: GOALS, METHODS, AND CURRICULUM

Dewey, J. (1976). The child and the curriculum. In J.A. Boydston (Ed.), *John Dewey: The middle works, 1899–1924* (Vol. 2, pp. 273–291). Carbondale: Southern Illinois University Press.

Ball, D.L. (1993). With an eye on the mathematical horizon: Dilemmas of teaching elementary school mathematics. *The Elementary School Journal, 93*(4), 373–397.

Ginsburg, H.P., & Amit, M. (2008). What is teaching mathematics to young children? A theoretical perspective and case study. *Journal of Applied Developmental Psychology, 29*(4), 274–285.

Balfanz, R., Ginsburg, H., & Greenes, C. (2003, January). The "Big Math for Little Kids" Early Childhood Mathematics Program. *Teaching Children Mathematics, 9*(5), 264–268.

12. NOVEMBER 20: TEACHING: MANIPULATIVES, GAMES, AND COMPUTER SOFTWARE

Stern, M.B. (2011). Multisensory mathematics instruction. In J. Birsh (Ed.), *Multisensory teaching of basic language skills* (3rd ed., pp. 457–479). Baltimore, MD: Paul H. Brookes Publishing Co. Retrieved from http://sternmath.com/about-the-program.html

Mix, K.S. (2010). Spatial tools for mathematical thought. In K.S. Mix, L.B. Smith, & M. Gasser (Eds.), *The spatial foundations of learning and cognition* (pp. 41–66). Oxford, England: Oxford University Press.

Moyer, P.S. (2001). Are we having fun yet? How teachers use manipulatives to teach mathematics. *Educational Studies in Mathematics, 46*, 175–197.

Siegler, R.S., & Ramani, G.B. (2009). Playing linear number board games—but not circular ones—improves low-income preschoolers' numerical understanding. *Journal of Educational Psychology, 101*(3), 545–560.

13. DECEMBER 4: FORMATIVE ASSESSMENT AND TEACHING: TWO SIDES OF THE SAME COIN

Ginsburg, H.P. (2009). The challenge of formative assessment in mathematics education: Children's minds, teachers' minds. *Human Development, 52*, 109–128.

Ginsburg, H.P., Pappas, P., Lee, Y.-S., & Chiong, C. (2011, November). How did you get that answer? Computer assessments of young children's mathematical minds. In P. Noyce (Ed.), *New frontiers in formative assessment* (pp. 49–67). Cambridge, MA: Harvard University Press.

14. DECEMBER 11: WRAP-UP

Class discussion of what has been learned about key issues.

Index

Figures, tables, and notes are indicated by the letter *f*, *t*, and *n*, respectively.

Accountability, 10, 12–14, 202
Accreditation Facilitation Projects, 14
Adaptive technologies, 133
Addition, 64–71, 131–133, 132*f*, 212, 214
Administration for Children and Families, 15, 101, 105
Algebra, 69–70, 79–80, 214–215
American Association of Colleges for Teacher Education, 126
American Federation of Teachers, 134
Annie E. Casey Foundation, 194
Apgar score, 1
Applied Problems Subtest of Woodcock Johnson-III, 147
Apps, 54, 133–134
Arithmetic, *see* Mathematics
Asia, 11, 59
Assessments, 12–14, 71
   authenticity in, 37
   clinical interviews, 53, 55, 58–59, 73*n*4, 74*n*16, 82, 128, 213
   comprehensive mathematics, 144
   conducting and using, 105–106
   formative, 60, 85–88, 121, 128, 191, 201–204, 215
   geometric sense and, 175, 178, 187–188
   in-class, 77, 85–88
   professional development and, 109
   using videos for, 178–179
At-risk learners, 89, 154–155
Australia, 10–11

Backward counting, 60
Base-10 system, 33, 58, 60
BB/TRIAD, *see* Building Blocks (BB) math curriculum; Technology-enhanced, Research-based Instruction, Assessment, and professional Development
Big Math for Little Kids, 20, 215
Blended learning, 127–128, 129*f*, 133, 145
   alternatives to, 129–130
   professional development through, 127–128, 127*f*, 128*f*, 186–188, 187*f*
Block play, 54, 82, 168–171
Board games, 54, 215
Brenneman, Kimberly, 149*t*

Building Blocks (BB) math curriculum, 20, 144, 174, 184–185, *see also* Technology-enhanced, Research-based Instruction, Assessment, and professional Development (TRIAD)

California, 6–7, 181, 182*t*
Cardinality, 34, 62–65
"Carrying" (in arithmetic), 59
CCSSO, *see* Council of Chief State School Officers
CDA, *see* Child Development Associate credentials
Center for Advanced Study of Teaching and Learning, 125, 149*t*
Center for New Media Teaching and Learning, 183, *see also* Video Interactions for Teaching and Learning (VITAL)
Certification, 3–5, 7–9, 24, 39, 137, 190
CGI, *see* Cognitively Guided Instruction
Chen, Jie-Qi, 149*t*
Chicago Public School District, 146–147, 149*t*
Chicago School Readiness Project, 102
Child Care and Development Fund, 11
Child care centers, 4, 7–8, 31, 43, 58
Child Development and Early Learning Framework, 19
Child Development Associate (CDA) credentials, 7, 9–10, 75
Child-centered education, 14, 37–38
Child–teacher interactions, 12–13, 83–85, 125–126, 178, 181, 186
China, 59
CK-12, 135
Classroom Assessment Scoring System® (CLASS™), 179–180, 182*t*
Classroom environment, 87–91
Classroom Observation of Early Mathematics Environment and Teaching (COEMET), 180, 185
Classroom observations, 85, 150, 180
Clements, Douglas, 149*t*
Clinical interviews, 53, 55, 58–59, 73*n*4, 74*n*16, 82, 128, 213
Coaching, 98–105, 110–112, 142–154, 149*t*, 157–159, 182*t*

217

COEMET, see Classroom Observation of Early Mathematics Environment and Teaching
Cognitively Guided Instruction (CGI), 86–87, 157–158
Colorado, 7
Columbia University, 126–128, 183, see also Video Interactions for Teaching and Learning (VITAL)
Committee on Early Childhood Mathematics, 35, 107
Common Core Library, 134
Common Core State Standards, xvii, 16, 19, 42, 44, 46, 106, 201, 208n2
  curricular resources for, 134–135
  edTPA and, 126
  on mathematics, 68–69
Common School, 2
Common Sense Media, 134
Computers, 20–21, 54, 185, 215
Connected Educators web site, 135
"Conservation of number" problem, 65
Continuity, 12, 14
Council for Exceptional Children, 134
Council of Chief State School Officers (CCSSO), 11, 42
Counting, 56–65, 212
  backward, 60
  base-10 system, 33, 58, 60
  objects, 34, 57, 61–66, 61f, 62f, 63f
  rote, 44
Cuisenaire rods and blocks, 20
Curricula, 20–21
  Building Blocks (BB) math, 20, 144, 174, 184–185
  Common Core State Standards resources, 134–135
  HighScope, 152
  implementation of, 89, 106
  interdisciplinary, 145
  Numbers Plus, 152
  professional development needs and, 151–156
Curriculum Focal Points, 19, 35
Curriki, 135

Daily calendar activities, 30, 33, 44, 90
Delaware, 7
Demographics, see Race; Socioeconomic status
"Designed Video," 121–123
Development of Mathematical Thinking course, 127–128, 127f
Developmentally appropriate instruction, 24, 32, 36–38, 72, 117, 201
Didactic teaching, 56

Differentiation, 167
DreamBox, 133
Dual-language learners, 12, 14, 18, 106, 155–156, 162, 167–171, see also SciMath-DLL
Dynamic memory theory, 178

Early Childhood Collaborative, 75
Early Childhood Data Collaborative, 13
Early childhood education (ECE)
  evolution of, 4–5
  overacademization of, 2–3
  policy/system building in, 11–12
  popularization of, 10–11
  professional development and, 1–28
  program quality in, 12–14
  "schoolification" of, 14
  trends in, 10–14
Early Childhood Educator Professional Development (ECEPD) programs, 98–105, 108
Early childhood educators (ECEs), xviii, 4–10
  beliefs about math, 33–38, 41, 45, 88, 188–191
  certification of, 3–5, 7–9, 24, 39, 137, 190
  child–teacher interactions, 12–13, 83–85, 125–126, 178, 181, 186
  compensation of, 6
  demographics among, 5–6
  gender of, 37, 78, 199
  instructional behaviors of, 77
  intermediary inventive mind of, 72
  mathematical knowledge of, 15–21, 33–38, 78, 190–191
  mathematics goals for, 75–95
  mathematics preparation of, 15, 21–22
  recruiting, 199–200
  requirements for, 7–9, 15, 76
  role of teacher as learner, 193
  supports for, 200
  teacher candidates, 126, 131–132
  turnover and instability among, 6–7
  typical profiles of, 30
  voluntary credentials of, 9–10
Early Childhood Faculty Connections, 136–137
Early Childhood Learning and Knowledge Center (ECLKC), 43
Early Childhood Longitudinal Study-Kindergarten (ECLS-K), 47n1, 47n4, 76
"Early Childhood Math Education Action Planning Tool," 46–47, 138

Early childhood mathematics education (ECME)
  change in, 10–24
  content of, 31–32, 71
  developmentally appropriate, 36–37
  diversity of students and, 37–38, 156
  importance of, 16
  improvements in, 18–21
  methods in, 32–33
  politics of, 207
  professional development and, 1–28
  research about, 17–18
  technology and, 33, 119–138
  time spent, 31
  typical, 30–33
  work force, xviii, see also Early childhood educators
Early Childhood Mathematics: Promoting Good Beginnings, 38–39
  see also SciMath-DLL
Early Childhood STEM Lab, 149t, 165
Early learning and development standards (ELDS), 13, 19
Early Learning Challenge Fund, 12–13, 106
Early Math Collaborative (EMC), 146–148, 149t
Early Mathematics Education Project, see Early Math Collaborative
ECE, see Early childhood education
ECEPD, see Early Childhood Educator Professional Development programs
ECEs, see Early childhood educators
ECLKC, see Early Childhood Learning and Knowledge Center
ECLS-K, see Early Childhood Longitudinal Study-Kindergarten
ECME, see Early childhood mathematics education
EdSurge, 134
edTPA, 126–127
Education Next, 1
"Effective Classroom Interactions: Supporting Young Children's Development," 129
ELDS, see Early learning and development standards
EMC, see Early Math Collaborative
Entering the Child's Mind: The Clinical Interview in Psychological Research and Practice, 82
Enumeration, 34, 57, 60–66, 61f, 62f, 63f
Environmental and lesson checklist, 88
Equal sign, 68–70, 73n14
Erikson Institute, 135, 149t
Ethnicity, 5–6, 37–38
Everyday math, 53–54, 68–70, 73n2, 76, 200, 212

Family child care homes, 4–5, 7–8, 31, 33, 43
Family Math project, 206
Female teachers, 37, 78, 199
Fidelity of Implementation measure, 185
Flickr, 127
Formal schooling, 2, 53
Formative assessments, 60, 85–88, 121, 128, 191, 201–204, 215
Foundations of Learning program, 102–103
Fred Rogers Center for Early Learning and Children's Media, 21
Free play, 30, 32, 123, 201–202, 212
Funding, 1–2, 12–13, 106, 147–148

Games
  as assessment tools, 86
  board, 54, 215
  Mistake game, 57–58
  Pirate Coins, 66–67
  Snake, 86
Gender, 37, 78, 199
Geometry, 3, 35, 47n3, 77–81, 86, 88, 214
  assessment of, 175, 178, 187–188
  shape recognition and, 35–37, 71, 80–81, 133, 144, 156, 214
  spatial reasoning and, 35, 77, 108, 147, 175, 214
  vocabulary of, 80–81
Gross enrollment rate (GER), 10

Hands-on learning experiences, 20
Hawaii, 7
Head Start, xviii, 8, 15, 42, 205
  Child Development and Early Learning Framework, 19
  Early Childhood Learning and Knowledge Center (ECLKC), 43
  Improving Head Start for School Readiness Act of 2007 (Head Start Act of 2007; PL 110-134), 8, 42–43
  professional development for teachers in, 42–43
Higher education, xiv, 9, 200
  degree programs, 9, 47n5, 110
  online degree programs, 23, 129–130
  successful professional development approaches and, 142–143
HighScope curriculum, 152

i3, see Investing in Innovation grant
IDEA, see Individuals with Disabilities Education Act

IES, *see* Institute of Education Sciences
IHEs, *see* Institutions of higher education
iMovie for Mac, 126
Improving Head Start for School Readiness Act of 2007 (Head Start Act of 2007; PL 110-134), 8, 42–43
Individuals with Disabilities Education Act (IDEA), 11
Institute of Education Sciences (IES), 146, 162n5
Institutions of higher education (IHEs), *see* Higher education
Instruction, *see* Teaching
Instructional behaviors, 77–78
Instructors, *see* Early childhood educators
Intentional teaching, 32, 71–73, 87–88, 174–181, 177f, 201–202
Interdisciplinary curricula, 145
Intermediary inventive minds, 73
Investing in Innovation (i3) grant, 147–148

James, William, 72
Joan Ganz Cooney Center, 131

Kaltura, 126–127
Khan Academy, 131–132, 132f
Khan, Sal, 131
#Kinderchat community, 135
Kindergarten
  education methods in, 33
  entry assessments, 13–14, 106, 109, 111–112
  math education in, 31–38
  standards in, 19
Kinzie, Mabel, 149t
"Knowledge-based" learning, 203

Lead teachers, 102–104, 108–109, 152–154
Low income, 8, 20, 30, 32, 98,
  see also Socioeconomic status

Manipulatives, 20–21, 37, 72, 84–85, 158, 215
Maryland, 12–13
Massachusetts, 7, 12
Massive Open Online Courses (MOOCs), 129
Math Matters, 17–18
Math methods courses, 30, 40, 47n5
Mathematical talk, 71, 154–155

Mathematical thinking, 214
  counting words, 56–60
  syllabus for a course on, 211–215
  typical early childhood, 53–74, 80–82
  see also Numbers
Mathematics
  arithmetic operators, 68–69
  "carrying" in, 59
  content knowledge, 34–35, 77–78
  everyday, 53–54, 68–70, 73n2, 76, 200, 212
  gender and, 37, 78, 199
  teacher attitudes and beliefs about, 33–38, 41, 45, 88, 188–191
  see also Mathematical talk; Mathematical thinking
Mathematics Belief Instrument (MBI), 188–189
Mathematics Knowledge for Teaching (MKT) Measures, 177–178, 182
Mathematics Learning in Early Childhood, xiii, 17
Mathematics Professional Development Institutes, 181, 182t
Mathematization of children's experiences, 84
Mathmateer, 133
MBI, *see* Mathematics Belief Instrument
Media Thread, 127, 182t, 183
Metacognition, 73n11, 214
Minnesota, 7
Minus sign, 68–69
Mistake game, 57–58
MKT, *see* Mathematics Knowledge for Teaching Measures
MOOCs, *see* Massive Open Online Courses
Mount Holyoke Mathematics Leadership Program, 154
M-TEAM standardized test, 86
MyTeachingPartner-Math/Science (MTP-M/S), 145–146, 149t, 150–153, 182t, 186–188

NACCRRA, *see* National Association of Child Care Resource and Referral Agencies
NAEYC, *see* National Association for the Education of Young Children
"Nation at Risk, A," 2
National Association for the Education of Young Children (NAEYC), 16, 19, 38–40, 71, 80, 119, 134, 175–176
National Association of Child Care Resource and Referral Agencies (NACCRRA), 43

National Board for Professional Teaching Standards (NBPTS), 9, 76
National Center for Early Development and Learning (NCEDL), 20, 31–32, 47n1
National Commission for Education and the Economy, 11
National Council for Accreditation of Teacher Education (NCATE), 40, 119
National Council of Teachers of Mathematics (NCTM), 16, 19, 35, 38–39, 47n1, 71, 80, 175–176
National Governors Association, 11, 42, see also Common Core State Standards
National Institute for Early Education Research, 149t, 165
National Mathematics Advisory Panel, 78, 88–89, 173, 190
National Research Council (NRC), xiii, xvii, 17, 24, 31–32, 47n1, 77, 142, 193
National Science Foundation (NSF), 89, 128, 154, 162n5
NBPTS, see National Board for Professional Teaching Standards
NCATE, see National Council for Accreditation of Teacher Education
NCEDL, see National Center for Early Development and Learning
NCTM, see National Council of Teachers of Mathematics
New Hampshire, 7–9
New Jersey, 7, 149t, 169–170
New York City Department of Education, 134
NRC, see National Research Council
NSF, see National Science Foundation
Number sense, 35, 108, 188
Numbers, 213–214
Numbers Plus curriculum, 152
Numerals, 54, 68–70, 168–171, 213

Ohio, 13, 129
"One, Two, Buckle My Shoe," 2–3
Online degree programs, 23, 129–130

Panda Path Early Learning Center, 75–76
Park Math, 133
Patterns, 35, 54, 79–80, 83, 123, 214–215
PBS, see Public Broadcasting Service
PD, see Professional development
Pedagogical Content Knowledge for Preschool Mathematics Interview, 34–35
Pennsylvania, 8–9, 12
Peripatetic method, 74n16
Personal learning networks (PLNs), 135

Photobucket, 126
Piaget, Ben, 65, 70–71, 121, 213
Pirate Coins game, 66–67
Planned variation studies, 188, 191–192
Play
  block, 54, 82, 168–171
  free, 30, 32, 123, 201–202, 212
  productive, 3
PL 108-446, 146, 162n5
PLCs, see Professional learning communities
PLNs, see Personal learning networks
Plus sign, 68–69
Poverty, 20, 118, see also Socioeconomic status
PowerDirector for PC, 126
"Practice-based" learning, 110, 157, 203
Praxis exam, 8
Prekindergarten, 5–8
  mathematics in, 20, 31–32, 47n1, 149t
  public, 6–8, 31, 75–76
  standards for, 19–20
  see also Curricula
Prekindergarten teachers, 8, 32, 90
Preparing Teachers: Building Evidence for Sound Policy, 40
Preschool, 10–11
  math education in, 31–38
  professional development for, 145, 167–171
Preschool teachers, 5–6, 18, 179
Problem-solving activities, 78, 158
Productive play, 3
Professional development (PD), xiii–xiv, xix, 1–51, 97f
  access to, 38, 153–154
  adequacy and authenticity in, 118–119, 136
  assessment information and, 109
  blended learning and, 127–128, 127f, 128f, 186–188, 187f
  challenges for, 109
  child characteristics and, 154–156
  coaching, 98–105, 110–112, 142–154, 149t, 157–159, 182t
  collective participation in, 102–104, 108–109
  curricular needs, 151–156
  direct focus on practice in, 108
  dosage and scale of, 104–105, 193–194
  effective, 97–115
  evaluating/assessing, xix–xx, 159–161, 173–198, 182t, 205–206
  future research needs in, 159–161, 189–193
  gap between preserve and in-service programs, 110–111

Professional development (PD)—*continued*
  group training and individualized coaching in, 102
  Head Start programs and, 42–43
  higher education and math content, 38–44, 200
  individualization of supports for adult learners, 153
  in-service teachers, 41–44, 203–205
  literature review, 103–104
  math-related field experiences and, 40, 202–203
  models, 173–174
  online communities and, 136–137
  organizational context and, 106, 109
  practice-focused, 100–102
  in preservice settings, 117–139, 141–171, 191
  recommendations for, 78, 82–83, 84–85, 87, 90–91, 175
  research-based models for, 143–145, 149*t*
  role of teacher as learner, 193
  specialists, 36
  state quality rating and improvement systems and, 43–44
  sustainable, 148–153
  teacher characteristics and, 188–189
  technology and, 117–139
  theories of change in, 174–176, 189–190
  time for, 148–156
  see also Early Math Collaborative (EMC); Higher education; MyTeachingPartner-Math/Science; Technology-enhanced, Research-based Instruction, Assessment, and professional Development (TRIAD)
Professional Development for Culturally Relevant Teaching and Learning in Pre-K mathematics, 150*t*, 157–158
Professional learning communities (PLCs), 89, 91, 151–152, 157, 159, 162
Project Pad, 127
"Promising practices," 77
Public Broadcasting Service (PBS), 206
Public engagement, 206–207

Quality Rating and Improvement System (QRIS), 9, 14, 43–44, 46, 101–103, 110–111
Quality Teaching Library, 182*t*, *see also* Classroom Assessment Scoring System
Quality Teaching Video Library, 145

Race, 5–6, 37–38
Race to the Top funding, 106
Reach Out and Read program, 206

Research to Action, 194
Richmond, Julius, 22
Rote learning, 36, 56, 154

Sarama, Julie, 149*t*
Scaffolding, 32, 44–45, 56, 59, 66, 84
School reform, 2
SciMath-DLL, 149*t*, 152, 154, 156–157, 167–171
Seeds for Success evaluation, 111
SES, *see* Socioeconomic status
Shape recognition, 35–37, 71, 80–81, 133, 144, 156, 214
Share My Lesson project, 134
Snake game, 86
Social learning theory, 178
Social media tools, 135
Socioeconomic status (SES), 37, 154–155
Socrates, 74*n*16
Spatial reasoning, 35, 77, 108, 147, 175, 214
Spirolaterals, 79–80, *see also* Algebra
Standardized tests, 56, 66, 86
Standards, 12–14, 19, 39, 81, *see also* Common Core State Standards; Early learning and development standards
Stanford University, 126
Start-from-one rule, 58
Storybooks, 54, 147, 206
Study of Instructional Improvement, 182
Subitizing, 73*n*8, 86, 156
Subtraction, 64–68, 212
Survey of Pedagogical Content Knowledge in Early Childhood Mathematics, 35, 47*n*4
Symbols, 68–70, 213, *see also* Numbers

Talks to Teachers, 72
T.E.A.C.H., *see* Teacher Education and Compensation Helps program
Teach for America, 8
Teacher Education and Compensation Helps (T.E.A.C.H.) program, 6
Teachers, *see* Early childhood educators
Teaching
  as an "art," 72
  child–teacher interactions, 12–13, 83–85, 125–126, 178, 181, 186
  class environments and, 87–91
  culturally relevant, 150*t*, 157–158
  developmentally appropriate instruction, 24, 32, 36–38, 72, 117, 201
  effective practices, 83–84
  intentional, 32, 71–73, 87–88, 174–181, 177*f*, 201–202
  see also Professional development

Teaching–learning paths, 36, 81–82, 88
TEAM, *see* Tools for Early Assessment in Math
Team Umizoomi, 133
Technology, xx, 20–21, 84–75, 138
  adaptive, 133
  early childhood mathematics education and, 33, 119–138
  professional development and, 117–139
  role of, 137–138
Technology-enhanced, Research-based Instruction, Assessment, and professional Development (TRIAD), 143–145, 149*t*, 152, 182*t*, 184–186, 192–193, *see also* Building blocks (BB) math curriculum
TEMA-3 standardized test, 86
Testing, 56, 66, 86
Theories of change, 174–176, 189–190, 194
Time To Know, 133
Tools for Early Assessment in Math (TEAM), 175
TRIAD, *see* Technology-enhanced, Research-based Instruction, Assessment, and professional Development
Twitter, 135, 151

United Nations Educational, Scientific and Cultural Organization (UNESCO), 10
University of Denver, 145, 149*t*
University of Virginia, 129, 149*t*
University of Wisconsin–Madison, 150*t*, 158
U.S. Bureau of Labor Statistics, 5
U.S. Department of Education, 98, 135, 147
U.S. Department of Health and Human Services, 105

VAIL, *see* Video Assessment of Interactions in Learning
Venn diagrams, 80
Verbal counting, 33, 144
Vermont, 7
Video Assessment of Interactions in Learning (VAIL), 179
Video Interactions for Teaching and Learning (VITAL), 125, 127–128, 129*f*, 182*t*, 183–184
Video-ANT, 127
Videos
  analyzing, 120–121
  annotating, 124–127
  assessments and, 178–179
  creating, 124–127
  "designed," 121–123
  teacher practice, 126
Vimeo, 127
Virtual learning communities, 136–137
VITAL, *see* Video Interactions for Teaching and Learning

Wager, Anita, 150*t*
Washington state, 12, 111
WeVideo, 126
"Whole child," xvii
Wisconsin, 150*t*
Woodcock Johnson-III, 147

Young children
  child–teacher interactions, 12–13, 83–85, 125–126, 178, 181, 186
  mathematical development of, xix, 53–74, 80–82
YouTube, 127, 131, 207